CAMBRIDGE STUDIES IN
MATHEMATICAL BIOLOGY: 4

Editors

C. CANNINGS

Department of Probability and Statistics, University of Sheffield

F. HOPPENSTEADT

Department of Mathematics, University of Utah

MATHEMATICAL METHODS OF POPULATION BIOLOGY

FRANK C. HOPPENSTEADT
Professor of Mathematics, University of Utah

Mathematical methods of population biology

CAMBRIDGE UNIVERSITY PRESS

Cambridge

London   New York   New Rochelle

Melbourne   Sydney

Published by the Press Syndicate of the University of Cambridge
The Pitt Building, Trumpington Street, Cambridge CB2 1RP
32 East 57th Street, New York, NY 10022, USA
296 Beaconsfield Parade, Middle Park, Melbourne 3206, Australia

First published 1982

Printed in the United States of America

Library of Congress Cataloging in Publication Data
Hoppensteadt, F. C.
Mathematical methods of population biology.
(Cambridge studies in mathematical
biology ; 4)
Bibliography: p.
Includes indexes.
1. Population biology – Mathematical models.
I. Title. II. Series. [DNLM: 1. Biometry.
2. Models, Biological. 3. Population.
QH 323.5 H798m]
QH352.H66   1981   574.5'248'0724   81-9977   AACR2
ISBN 0 521 23846 3 hard covers
ISBN 0 521 28256 X paperback

CONTENTS

PREFACE

This book is intended as an introduction to methods that are useful for studying population phenomena. The models are presented in terms of difference equations. Experience has shown that this approach facilitates communicating the derivation of models and statements of results about them to scientists who do not have a strong mathematical background. However, in most cases of difference equations the mathematician must exert greater analytical effort because many of the features of calculus are not available in this setting. Important models that do involve extensive use of calculus are presented in exercises. In most cases, the exercises are fronts for presenting models more detailed than those derived and studied in the text.

The material is graded in terms of mathematical difficulty. The earlier chapters involve elementary difference equations, and later chapters involve topics requiring more mathematical preparation. First, models of total population and population age structure are derived and studied. Next, models of random population events are presented in terms of Markov chains. The final two chapters deal with mathematical methods used to uncover qualitative behavior of more complicated difference equations. For example, the material on geographically distributed populations eventually involves nonlinear diffusion equations. In each case, the chapter begins with a simple model, usually of some historical interest, that motivates the primary goals of the chapter.

The approach taken here evolved over many years from sets of lectures presented at New York University, the Courant Institute of Mathematical Sciences, and the University of Utah, and many students and colleagues participated in and contributed to the topics covered. In particular, I thank J. B. Keller for his insights and continued support throughout the development of the early aspects of this work.

I also thank the many people who helped prepare the manuscripts during this development period. Among these are A. Costadasi, J. Figueroa, C. Engle, and J. Hadley.

F. C. Hoppensteadt

Salt Lake City
September 1981

1

Population dynamics

1.1 Iteration and parameter estimation

Many insects hatch from eggs, mature, reproduce as mature adults, and then die. Frequently, this life cycle is synchronized over an entire population. Populations of such organisms can be studied by counting the number of adults at reproduction time. This results in a sequence of numbers, say P_1, P_2, P_3, P_4, ..., where P_n is the number of adults at the nth reproduction time. A list of numbers like this can be studied in various ways to predict the changes such a population will make over many generations in the future and to characterize such data by a few parameters that will enable easy comparison between populations raised in alternate environments and comparison among species.

One might expect there to be a functional relation between successive generations:

$$P_{n+1} = f(P_n)$$

f is called the *reproduction function*. A simple test of this is to plot the pairs of points (P_n, P_{n+1}), $n = 1, 2, \ldots$, on graph paper. Surprisingly, it is often found that these data actually lie on a smooth curve, called the population's *reproduction curve*.

A. *Malthus's model*

Figure 1.1 depicts a hypothetical case where the data lie approximately on a straight line. A straight line can be fit to the data in Figure 1.1 using the least-squares method (see the Appendix and Exercise 1.1), say with the result that

$$P_{n+1} = rP_n \qquad \text{(model)} \tag{1}$$

where r is determined from the least-squares fit. The parameter r

1

characterizes the data set. It is the slope of the straight-line fit to the observed data, and the line $y = rx$ is the reproduction curve for the data.

Algebraic iteration. Using the model, we can determine theoretical values P_n in terms of n, r, and the initial size P_1 by successive back substitutions:

$$P_{n+1} = rP_n = r(rP_{n-1}) = r^2 P_{n-1}$$
$$= r^2(rP_{n-2}) = r^3 P_{n-2} = \ldots = r^n P_1$$

Thus, a population having a straight line as a reproduction curve grows or dies like a geometric sequence. If $r > 1$, $P_n \nearrow \infty$ as $n \to \infty$, and if $r < 1$, $P_n \searrow 0$ as $n \to \infty$.

Unfortunately, it is not possible to solve explicitly most of the models encountered in practice. So another method proves to be useful for determining qualitative behavior of the population numbers. It is based on the reproduction curve.

Geometric iteration. First, we plot the theoretical reproduction curve and a one-to-one reference line on the $P_n P_{n+1}$-axes. Then we start with P_1 on the x-axis, evaluate P_2 on the reproduction curve, and reflect this value back to the horizontal axis through the one-to-one line as

Figure 1.1. Population data, plotted pairwise.

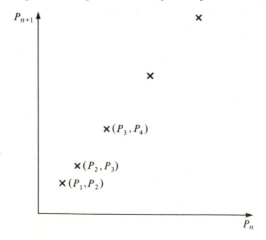

shown in Figure 1.2*a*. This procedure can be repeated as shown in Figure 1.2*b*. This method is sometimes referred to as *cobwebbing*. It shows that if $r > 1$, then the sequence of population sizes increases without bound.

The theoretical model

$$P_{n+1} = rP_n$$

is the one introduced by Malthus (1798). He argued that human populations grow geometrically, but that their food resources grow algebraically (i.e., proportional to n). This eventually leads to a limiting of population growth, which is not accounted for in the model (1). If limiting occurs, it must be that r changes with population size, approaching zero as the population gets large.

B. *Verhulst's model*

Populations whose growth is limited by a resource, such as nutrient or living space, may have data $P_1, P_2, P_3, P_4, \ldots$, as depicted in Figure 1.3. A reproduction curve can be found by plotting the points (P_n, P_{n+1}) as before, but there are many possible functional relations that give the same qualitative behavior. The simplest one, in which r approaches zero as population size increases, is given by

$$f(P) = \frac{\rho P}{P + K}$$

Figure 1.2. (*a*) Iteration of Malthus's model, $r > 1$. (*b*) Cobwebbing method.

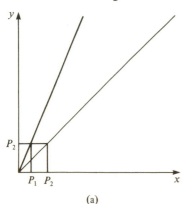

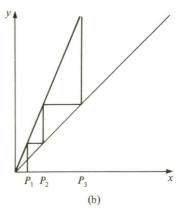

(a)   (b)

where ρ and K are parameters. The corresponding model is

$$P_{n+1} = \frac{\rho P_n}{P_n + K} \tag{2}$$

This is a discrete version of a model proposed by Verhulst in 1845, and it can be analyzed as Malthus's model was.

Algebraic iteration. Setting $Q_n = 1/P_n$, we convert the Verhulst model into

$$Q_{n+1} = \frac{1}{\rho} + \frac{K}{\rho} Q_n$$

Q_n (hence P_n) can be determined in terms of n, ρ, K, and Q_1 ($= 1/P_1$), by successive back substitution:

$$
\begin{aligned}
Q_{n+1} &= \frac{1}{\rho} + \left(\frac{K}{\rho}\right) Q_n = \frac{1}{\rho} + \frac{K}{\rho}\left(\frac{1}{\rho} + \left(\frac{K}{\rho}\right) Q_{n-1}\right) \\
&= \frac{1}{\rho} + \frac{K}{\rho}\left(\frac{1}{\rho}\right) + \left(\frac{K}{\rho}\right)^2 Q_{n-1} \\
&= \frac{1}{\rho}\left(1 + \frac{K}{\rho} + \ldots + \left(\frac{K}{\rho}\right)^{n-1}\right) + \left(\frac{K}{\rho}\right)^n Q_1 \\
&= \frac{1}{\rho}\left(\frac{1 - (K/\rho)^n}{1 - K/\rho}\right) + \left(\frac{K}{\rho}\right)^n Q_1
\end{aligned}
$$

Figure 1.3. Population data describing the approach to the carrying capacity.

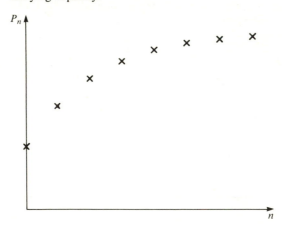

If $K/\rho < 1$, then $Q_n \to 1/(\rho - K)$, so $P_n \to \rho - K$. If $K/\rho = 1$, then $Q_n = Q_1 + n/\rho$, which grows algebraically; and if $K/\rho > 1$, then $Q_n \to \infty$, geometrically. In the last two cases $P_n \to 0$.

Because there is a linear relation between Q_n and Q_{n+1}, we can again estimate the model's parameters by a least-squares approximation. This shows that the parameters K and ρ in the Verhulst model can be determined by a standard straight-line approximation to the points $(1/P_n, 1/P_{n+1})$ and that the behavior of the population can be determined explicitly:

$$P_n \to \rho - K \quad \text{if} \quad K < \rho$$
$$P_n \to 0 \qquad \text{if} \quad K \geq \rho$$

This qualitative behavior can also be determined by cobwebbing.

Geometric iteration. The two cases $K < \rho$ and $K \geq \rho$ are depicted in Figure 1.4. Figure 1.4a shows the Verhulst reproduction curve when $\rho > K$. K represents the population size at which one half of the maximum possible offspring production in the population is realized. Cobwebbing in Figure 1.4a and b shows that P_n approaches $\rho - K$ and Q_n approaches $1/(\rho - K)$ in successive generations. Cobwebbing in Figure 1.4c and d shows that $P_n \to 0$ and $Q_n \to \infty$ in successive generations.

The Verhulst model was introduced to studies of fish populations by Beverton and Holt (1957), and we shall encounter this model again in Section 1.3.

C. *The predator pit*

If a population is reduced to a low level, predators can completely eliminate it. Other problems, such as difficulty in finding mates, also arise. For such reasons it is important to be able to account for possible extinction of the population. A reproduction model that does this is

$$P_{n+1} = \frac{\rho P_n^2}{(K + P_n^2)} \tag{3}$$

the corresponding reproduction curve is shown in Figure 1.5, where θ denotes the threshold of extinction: If $P_1 < \theta$, then $P_n \to 0$. Holling

(1965) has referred to the interval $0 < x < \theta$ as being the *predator pit*. If $P_1 > \theta$, then $P_n \to P^* = (\rho + (\rho^2 - 4K)^{1/2})/2$ as $n \to \infty$.

The parameters ρ and K can be estimated from data by using the least-squares method. First, let $Q_n = 1/P_n$. With this change the model becomes

$$Q_{n+1} = (K/\rho)Q_n^2 + (1/\rho)$$

The theory, therefore, predicts a parabolic relationship,

$$y = K/\rho \ x^2 + 1/\rho$$

between reciprocal population sizes in succeeding generations.

Figure 1.4. (*a*) Verhulst's model, $K < \rho$. (*b*) $Q_n = 1/P_n$, $K < \rho$. (*c*) $K > \rho$. (*d*) $K > \rho$.

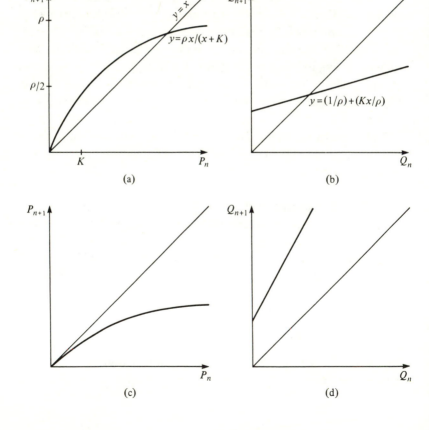

The use of least squares to find K/ρ and $1/\rho$ from data is illustrated in Exercise 1.2.

D. *Chaos*

The Malthus and Verhulst models seem to indicate that a population's dynamics can be determined, at least qualitatively, by geometric iteration: Observed data can be plotted to suggest an appropriate reproduction curve, and then cobwebbing can be used to describe the evolution of population size. Unfortunately, this is misleading. A slight modification of the Verhulst model leads to quite difficult problems.

If large numbers of adults can seriously deplete the environment of nutrients vital to survival of offspring, for example by defoliating and killing vegetation needed by offspring, quite complicated dynamics can occur.

This is illustrated by a model that has been extensively studied, although this "simple" model is still not completely understood. The reproduction model is

$$P_{n+1} = \rho P_n (K - P_n)_+$$

where $(K - P_n)_+$ denotes the positive part of $K - P_n$; that is,

Figure 1.5. Depensatory reproduction curve. $0 < P < \theta$ is referred to as the predator pit.

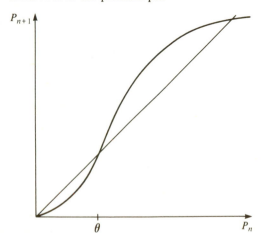

$$(K - P_n)_+ = K - P_n \quad \text{if} \quad P_n < K$$
$$= 0 \quad\quad \text{if} \quad P_n \geq K$$

The corresponding reproduction curve is depicted in Figure 1.6. The dynamics exhibited by this reproduction model depend sensitively on $K\rho$. For $K\rho < 1$, $P_n \to 0$; for $1 < K\rho < 3$, P_n approaches the static state $P^* = K - (1/\rho)$. For values of $K\rho$ greater than 3, the sequence P_n may be periodic or it may be chaotic. For example, if $3 < K\rho <$

Figure 1.6. Overcompensation reproduction curve.

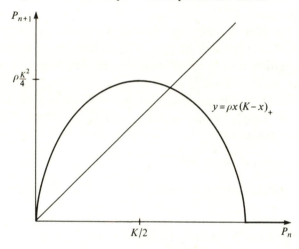

Figure 1.7. Bifurcation diagram for the overcompensation model.

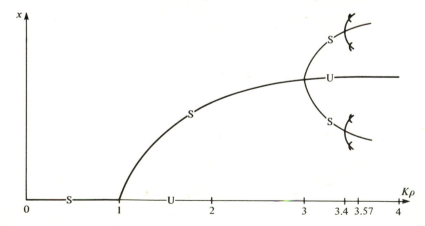

3.44, P_n approaches periodic form, where $P_{2n} = P_u^*$, $P_{2n+1} = P_l^*$. The results can be summarized by a bifurcation diagram as shown in Figure 1.7.

Figure 1.7 shows that for $K\rho < 1$, $x = 0$ is stable. As $K\rho$ increases through the value 1, $x = 0$ becomes an unstable static state and the new state $P = K - 1/\rho$ appears as a stable state: A bifurcation of static states occurs accompanied by an exchange of stabilities as $K\rho$ is increased through 1. At $K\rho = 3$, a cascade of bifurcations begins. At each, a periodic solution having twice the period of the previous one appears and there is an exchange of stabilities. Infinitely many such bifurcations have occurred by the time $K\rho$ reaches 3.58. Beyond this value, the solutions behave chaotically. Exercise 1.3 develops the behavior at $K\rho = 4$.

A similar model was introduced by Ricker (1956) and studied by May (1973), Oster, Hyman (1977), and others. The reproduction function in their studies has the form $P_{n+1} = rP_n \exp(-qP_n)$, where r and q are parameters. This model has qualitative behavior like the quadratic one described here. This sequence of bifurcation was described by Šarkovski (1964) and by Thom (1975).

1.2 Synchronization of populations

The model in Figure 1.6 shows that populations can have mechanisms that cause oscillations of population size numbers. Other things can also lead to oscillations. Following are two interesting examples that illustrate the phenomenon of population synchronization.

A. *Two-year salmon: synchronization by age structure*

Some species of Pacific salmon have a two-year life cycle. Eggs hatch in a river after spawning. After one year, the survivors are immature young, and after a second year they are mature adults returning to the river ready for reproduction. After reproduction, the adults die.

Let Y_n = number of immature young in year n, and A_n = number of reproducing adults in year n. As before, one expects that there is a functional relationship between the population age structure in year $n + 1$, (A_{n+1}, Y_{n+1}), and (A_n, Y_n), say

$$A_{n+1} = g(Y_n), \qquad Y_{n+1} = f(A_n)$$

Typical assumptions for this system are that reproduction is described by a Verhulst model with a predator threshold and that a certain proportion of young survive the year to maturity, say $g(Y) = \sigma Y$, where σ ($\sigma < 1$) measures the survival probability of young.

This model can be studied by geometric iteration. Recall that the cobwebbing method required only the horizontal axis, the one-to-one reference curve, and the reproduction curve. Therefore, we can describe the two age classes by plotting one on the horizontal and one

Figure 1.8. (*a*) $Y = f(A)$ depicts the reproduction curve. $A = \sigma Y$ depicts the survival curve of young maturing to adults. (*b*) Population distribution (A_1, Y_1) determines the distribution (A_2, Y_2). A_2 is the reflection of Y_1 in the survival curve; Y_2 is the reflection of A_1 in the reproduction curve; (*c*) Region I, balanced population distribution; Region II, extinction; Regions III and III', synchronized distributions.

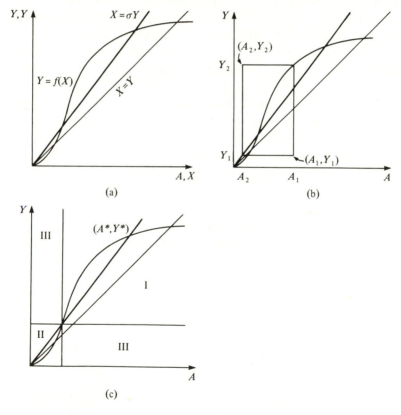

on the vertical axis. This is shown in Figure 1.8. The population in year 1 is described by the numbers Y_1, A_1; these determine a point with coordinates (A_1, Y_1) in Figure 1.8a. In the next year, the numbers become $Y_2 = f(A_1)$, $A_2 = \sigma Y_1$. The point (A_2, Y_2) can be determined geometrically as shown in Figure 1.8b. Geometric iteration shows that there are four regions, as shown in Figure 1.8c.

Region I. If (A_1, Y_1) is in region I, then the points (A_n, Y_n) approach the static state (A^*, Y^*) in successive generations. The static state is found by solving the equation

$$Y^* = f(A^*), \qquad A^* = \sigma Y^*$$

There are three solutions of this system; (A^*, Y^*) is the largest of these.

Region II. If (A_1, Y_1) is in region II, the points (A_n, Y_n) approach (0, 0). The population becomes extinct.

Regions III and III'. If (A_1, Y_1) is in III, then the even-year offspring (A_{2n}, Y_{2n}) approach $(0, Y^*)$ (i.e., $A_{2n} \to 0$ and $Y_{2n} \to Y^*$) and the odd year offspring (A_{2n+1}, Y_{2n+1}) approach $(A^*, 0)$. Similar behavior results for populations initially in III'. Table 1.1 shows the approximate population distribution after many years (n large).

The state $(A^*, Y,^*)$ resulting from region I is called the *balanced state*. Region I is its domain of attraction. Region II is the domain of attraction of the *extinct state* (0, 0). Regions III and III' are the domains of attraction of the *synchronized state* $\{(A^*, 0), (0, Y^*)\}$.

Table 1.1. *Approximate population distribution in regions III and III'*

Population		Year			
		$2n$	$2n + 1$	$2n + 2$	$2n + 3$
(A_1, Y_1) in III	Adults	A^*	0	A^*	0
	Young	0	Y^*	0	Y^*
(A_1, Y_1) in III'	Adults	0	A^*	0	A^*
	Young	Y^*	0	Y^*	0

B. *Synchronization of periodical insect emergences*

Various species of cicada (*Magicicada* spp.) have different emergence patterns. Those having life spans of seven years or less have emergences of comparable size every year. Cicadas having life spans of 13 and 17 years emerge in a synchronized way: There are large emergences every 13 and 17 years, respectively, but none in between. The length of life cycle alone does not explain this phenomenon.

Synchronization can occur when there is a limited carrying capacity for the environment, when there is a predator threshold that increases in response to a large emergence, or when these act together. A model will be derived for the general case, such as that by Hoppensteadt and Keller (1976), but each mechanism will be illustrated separately.

Adult cicadas mate and lay eggs in trees during a two-week time span, after which they die. The eggs drop to the ground, and hatch as nymphs. These enter the ground and attach to rootlets, where they reside until it is time to emerge. This time underground is almost the entire length of the life span. During this time, the nymphs pass through several instars. It is remarkable that at emergence time, nearly all of the surviving nymphs emerge simultaneously (Simons, 1979).

The nymphs emerge as mature adults, which then fly to nearby trees to repeat the cycle.

Consider a species having a life span of L years with reproduction occurring in year L followed by death of the parents. Let $X_{n-L} = $ number of nymphs becoming established underground in year $n - L$, and $\alpha = $ the survival rate of nymphs underground per year. Then $X_{n-L}\alpha^L$ of them survive L years and emerge as adults in year n. When they emerge, predators will eliminate as many as P_n of them. There will be none left for mating if $X_{n-L}\alpha^L \leq P_n$; otherwise, $X_{n-L}\alpha^L - P_n$ survive to reproduce. Therefore,

$$(X_{n-L}\alpha^L - P_n)_+ = \text{number of adults emerging in year } n$$
$$\text{that survive predation}$$

Let $f = $ number of hatched nymphs becoming established underground that each adult produces in a breeding period. Then

$$H_n = \text{number of nymphs produced in year } n$$
$$= f(X_{n-L}\alpha^L - P_n)_+$$

These nymphs enter the ground if the residual carrying capacity K_n can contain them, that is, if $K_n \geq H_n$. If not, only enough nymphs enter the ground to fill the available space underground. The residual carrying capacity can be calculated as

$$K_n = \left(K - \sum_{a=1}^{L-1} X_{n-a}\alpha^a \right)_+,$$

where K is the total carrying capacity.

The nymphs becoming established in year n are

$$
\begin{aligned}
X_n &= \min(H_n, K_n) \\
&= \min\left(f(X_{n-L}\alpha^L - P_n)_+, \left(K - \sum_{a=1}^{L-1} X_{n-a}\alpha^a \right)_+ \right)
\end{aligned}
$$

The predation threshold P_n depends on the previous year's level and on the size of the previous year's emergence:

$$P_n = RP_{n-1} + AX_{n-1-L}\alpha^L$$

where R is the relaxation rate of predators and A measures the increment due to the preceding year's emergence, $X_{n-1-L}\alpha^L$. If the predation level remains constant from year to year, then $A = 0$, $R = 1$, so $P_n = P_{n-1}$. The equations for X_n and P_n make up the general model. The corresponding reproduction curve is shown in Figure 1.9.

If an emergence falls below P_n^*, their progeny approach extinction. Also, the full population approaches extinction unless $f\alpha^L > 1$ and $R < 1$, if $A \neq 0$. If there is no emergence in year n, there will be no emergence in the later years $n + NL$ for $N = 1, 2, \ldots$.

$L = pq$. Suppose that L can be written as the product of two integers p and q. We look for a stable synchronized solution having period p:

$$
\begin{aligned}
&X_{np} = X, \ X_{np+k} = 0 \quad \text{for} \quad k = 1, 2, \ldots, p - 1 \\
&P_{np} = P
\end{aligned}
$$

To ensure the solution is stable, we must have

$$X = K - \sum_{a=1}^{L-1} X_{n-a}\alpha^a = K - \sum_{k=1}^{q-1} X\alpha^{kp} \quad \text{or}$$

$$X = \frac{K(1 - \alpha^p)}{1 - \alpha^L}$$

Also,

$$P_{np+k} = P_k = \frac{R^{k-1}}{1 - R^p} \frac{KA\alpha^L(1 - \alpha^p)}{1 - \alpha^L}$$

It is necessary for stability that

$$X_n > P^*_{np} = \frac{fP_{np}}{f\alpha^L - 1}$$

Combining these formulas leads to the condition

(stability of *p*-period emergence, general case) $$S_p \equiv \frac{Af\alpha^L}{f\alpha^L - 1} \frac{R^{p-1}}{1 - R^p} < 1 \qquad (4)$$

for the existence of a stable solution. In the case of constant predation level, $A = 0$ and $R = 1$, and the stability condition becomes

(stability of *p*-period emergence; P = constant) $$S_p \equiv \frac{P(1 - \alpha^L)}{K(1 - \alpha^p)(f\alpha^L - 1)} < 1$$

The case $K_n = K$ illustrates that synchronization can occur from the

Figure 1.9. Cicada reproduction curve for a cohort established in the year $n - L$.

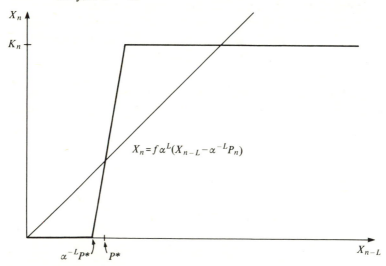

$$X_n = f\alpha^L(X_{n-L} - \alpha^{-L}P_n)$$

predation response alone. In this case the stability condition becomes

(stability of p-period
emergence;
K_n = constant)
$$S_p \equiv \frac{A\alpha^L}{K(f\alpha^L - 1)} \frac{R^{p-1}}{1 - R^p} < 1$$

In all these cases, S_p is a decreasing function of p. To see this in the first and third cases, where $R < 1$, we observe that the derivative of $F(x) = R^{x-1}/(1 - R^x) = e^{(x-1)\ln R}/(1 - e^{x\ln R})$ with respect to x is negative:

$$\frac{dF}{dx}(x)\bigg|_{x=a} = \frac{(\ln R)e^{(x-1)\ln R}}{(1 - e^{x\ln R})^2} < 0$$

For example, if $L = 12$, then $p = 1, 2, 3, 4, 6, 12$ are possible choices. It follows that in all three cases

$$S_1 > S_2 > S_3 > S_4 > S_6 > S_{12}.$$

Synchronization will occur if $S_1 > 1$, so there is no stable balanced solution, whereas $S_p < 1$ for some $p > 1$, so there is a stable p-period emergence. If L is a prime number, then since $S_1 > S_L$ it is possible for there to be an L-year emergence pattern where adults would be seen only every L years, whereas there is no stable balanced emergence pattern where comparable numbers would be seen every year.

For example, in case $P = 10^4$ ($=$ constant), we have the stability condition as shown in Table 1.2. Here "balanced initial data" refers to an experiment where an equal (sufficiently large) number of

Table 1.2. *Stability condition $K = 10^4$, $\alpha = .99$, $f = 10$*

		p		
L	1		L	Comment
7	$S_1 = .84$		$S_7 = .12$	Balanced initial data imply balanced emergence
13	$S_1 = 1.5$		$S_{13} = .13$	Balanced initial data imply synchronized emergence
17	$S_1 = 2.16$		$S_{17} = .13$	Balanced initial data imply synchronized emergence

nymphs is buried in each of L consecutive years to start the process. This gives initial data in region I in analogy with the preceding section (see Exercise 1.5).

These calculations of stable emergence patterns for cicadas show that synchronization can result from a limited carrying capacity, from a responding predation threshold, or from a combination of the two.

May beetles (*Melolontha* spp.) have life cycles of three and five years, depending on species. These are observed to have synchronized emergences. The oak eggar *(Lasi campa quercus)* has a one- or two-year cycle, depending on species. The two-year moth is periodic. Bamboo flowering is also synchronized. Its life span is approximately 120 years!

1.3   Exploitation of biological populations: fisheries

Reproduction models have been widely used in the management of exhaustible resources. The models are used to evaluate various management programs that might be used, usually with the goal of optimizing some economic performance index.

Fishery management provides good examples of the uses of mathematical theories to study the effects of optimization on populations. Mathematical theories have been used for a long time in the study of fisheries, primarily because of the inaccessibility of fish populations. The history of these applications and further topics in fishery management are described by Beverton and Holt (1957) and Clark (1976).

A fish population's dynamics can be described in rough terms by a stock-recruitment relationship, $R = f(S)$, where R denotes the number of fish introduced to the population (prior to harvesting) as offspring of a standing stock of size S. For simplicity it is assumed that the population's dynamics are described in terms of this function in the following way. If P_n denotes the stock size after harvesting in year n, then the next year's parent population will be

$$P_{n+1} = f(P_n) - h_n$$

where h_n denotes the harvest taken from the recruit population.

A. *Maximum sustained biological yield*
When the fishery is at equilibrium, $P = P_{equi}$, the formula

$$h_{\text{equi}} = f(P_{\text{equi}}) - P_{\text{equi}}$$

gives the biological yield of the fishery. This yield is maximized by a population maintained at the level P_{MSY} determined by the equation

$$(df/dP)(P_{\text{MSY}}) = 1$$

The corresponding yield,

$$Y_{\text{MSY}} = f(P_{\text{MSY}}) - P_{\text{MSY}}$$

is called the *maximum sustained yield* (MSY) of the population.

One technique used by fishery managers to control a fishery is to maintain it at MSY, although as we see later, this is a controversial method. The actual fish population is very difficult, often impossible, to monitor, and the only data a manager has to work with are the yield Y (say in tons of fish reported to be landed per week) and the effort E expended in harvesting. Effort might be measured in terms of standard fishing-boat days spent on the fishing grounds, where a standard fishing boat is taken appropriate to the particular fishery.

Therefore, it is necessary to reformulate the equilibrium problem in terms of the observable variables Y and E. First, an assumption is made about the relation between effort and corresponding harvest.

Effort–harvest relation. A unit effort produces a harvest of size qP from a population of size P, where q is called the coefficient of catchability and is independent of P, h, and E.

In the equilibrium case, the effort required to give a sustained yield $Y = f(P) - P$ is

$$E = (1/q) \sum_{k=P}^{f(P)} (1/k)$$
$$\doteq (1/q) \int_{P}^{f(P)} (dp/p) = (1/q) \log (f(P)/P)$$

since the effort required to reduce the population one unit from a population of size P is $1/qP$. The error made in this approximation is no greater than

$$(1/q) \left(\frac{1}{P} + \frac{1}{f(P)} \right)$$

The two equations

$$E = (1/q) \log (f(P)/P)$$
$$Y = f(P) - P$$

give an implicit relation between E and Y, parameterized by P. In special cases, P can be eliminated from these equations. For example, this can be done for a Beverton–Holt fishery for which

$$f(P) = bP/(P + a)$$

where $0 < a < b$. In this case,

$$E = (1/q) \log (b/(P + a)) \qquad Y = P(b - P - a)/(P + a)$$

so

$$Y = (be^{-qE} - a)(e^{qE} - 1)$$

This is plotted in Figure 1.10. The curve in Figure 1.10 is used in fishery management in the following way. When an increase in effort results in a decline in yield, the fishery exceeds MSY and effort is reduced, allowing the fishery to equilibrate. However, MSY is a bio-

Figure 1.10. Effort–yield curve for the Beverton–Holt fishery. E_{MSY} is the effort needed to produce the maximum sustained yield (MSY).

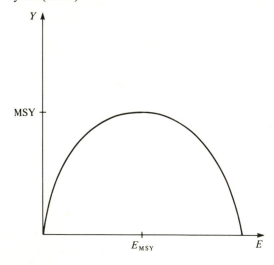

logical concept, not an economic one, so this management policy can lead to serious errors.

B. *Economic yield: bionomic equilibrium*

There are important differences between biological yield and economic yield of a fishery. These can be illustrated through a simple economic example. If p is the price per unit harvest received for fish and c is the cost per unit effort expended in harvesting, the revenue generated by the fishery is $R = ph - cE$. We suppose that p and c are constant over several generations. Although this is an unrealistic assumption, it is convenient. Recall that

$$E = (1/q) \log [f(P)/(f(P) - h)] \sim h/qf(P)$$

if $h \ll f(P)$. Thus, the current revenue is

$$R_n = [p - (c/qf(P_n))] h_n$$

In order to compare the economic cases with MSY, let us consider the equilibrium problem for *maximum current revenue* (MCR),

$$P = f(P) - h, \qquad R = ph - (c/qf(P))h = \text{maximum}$$

This problem is easily solved since the solution entails only finding the values of P, say P_{MCR}, which maximize

$$R = [p - (c/qf(P))](f(P) - P)$$

It can be shown for a Beverton–Holt fishery that $P_{MCR} > P_{MSY}$.

Another informative calculation shows the relation between E_{MSY} and the effort needed to sustain P_{MCR}, call it E_{MCR}. In fact, R may be written as $R = pY - cE$, so the revenue curve can easily be plotted on the effort–yield diagram as shown in Figure 1.11 for the Beverton–Holt fishery. Thus, $E_{MCR} < E_{MSY}$ and $P_{MCR} > P_{MSY}$.

C. *Maximum (dynamic) economic yield*

Equilibrium problems do not provide answers to questions of how transients influence optimal management policies or how robust these policies are with respect to errors. More important, the biological yield and the current revenue are probably not good performance indices for this economic system. To analyze these questions, we introduce a more realistic optimization model, which accounts for future revenues of the fishery. That is, the current harvest is compared with future values of money by considering a discounted total revenue of the fishery. One question this approach answers is whether or not the harvest should be increased now, for example to the limit of entirely eliminating the resource with the income being invested at current interest rates, to optimize the overall income from the resource. In order to address this question, a discount factor $\delta = 1/(1 + i)$, where i is the annual interest, is introduced, and the present value of future harvests is considered rather than the current revenue. The mathematical formulation of this optimization problem is

$$P_{n+1} = f(P_n) - h_n$$
$$PV[h] = \sum_{k=0}^{\infty} \delta^k(ph_k - cE_k) = \max \qquad \text{(discounted future revenue)}$$

Here $PV[h]$ gives the present value of the harvesting strategy that is given by the sequence $\{h_n\}$. The problem is to determine this harvesting strategy so as to optimize PV.

When $\delta = 1$, it can be shown (see Clark, 1976) that the optimal harvest strategy is to immediately reduce the population to P_{MCR} and then maintain equilibrium harvest h_{MCR} thereafter. Although this

Figure 1.11. Effort–yield curve ($Y = Y(E)$) and revenue curve ($Y = R(E)$) for the Beverton–Holt fishery. E_{MCR} is the effort needed to produce the maximum current revenue. Note that $E_{MCR} < E_{MSY}$.

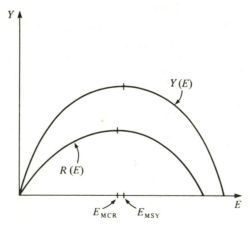

strategy may not be optimal when $c \neq 0$, numerical evidence indicates that it is nearly optimal in many cases. No attempt is made here to analyze this problem further.

The case $\delta = 0$ corresponds to an open-access fishery in which all revenue is dissipated. In this each fisherman totally discounts future harvests; if there is any revenue to be generated, someone will exert the effort. The mathematical formulation of this problem is

$$P_{n+1} = f(P_n) - h_n$$
$$R_n = ph_n - cE_n = 0 \qquad \text{(open-access fishery)}$$

Equilibria are given by $P = f(P)$, $h = 0$, and by

$$f(P^*) = c/pq, \qquad h^* = f(P^*) - P^*$$

In the first, no revenue is generated, so the second is the relevant one. The solution is depicted in Figure 1.12 for two cases. Figure 1.12 shows that an open-access fishery can collapse. Here P^* is less than the threshold of extinction.

The solution of the present-value problem lies between the two extreme cases of $\delta = 0$ and $\delta = 1$.

D. *Input–output models: multispecies fisheries*

In practice it is difficult to obtain enough data to reliably determine stock-recruitment relationships, especially when several

Figure 1.12. Open access fisheries. (*a*) Beverton–Holt fishery. (*b*) Depensatory fishery. This shows that an open-access fishery can lead to collapse of the fish population.

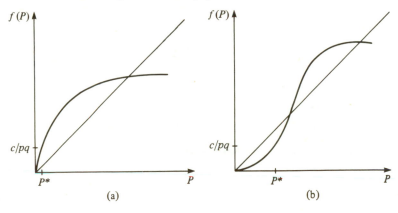

(a)   (b)

interacting species are to be investigated. Because of this we have
applied to fisheries methods that have been successfully used to analyze large, complex economics systems (Hoppensteadt and Sohn,
1979, 1980). First, some economic background on the input–output
method is given and then an application to the cod–haddock fishery
is briefly described.

Input–Output economics. An economy can be split into sectors (for
example, steel, coal, oil, iron ore) denoted by x_i. The vector

$$\mathbf{x} = \begin{bmatrix} x_1 \\ \cdot \\ \cdot \\ \cdot \\ x_n \end{bmatrix}$$

is called the *output vector*. Exogenous demands for the sectors are
given by a *demand vector* **d**. These are made up of the four components of the gross national product: personal consumption, investment,
government expenditures, and net exports. Finally, a flow matrix (or
technology matrix) describes the amounts of outputs consumed inter-
nally by the economy: a_{ij} = number of units of sector i needed to
produce a unit of sector j. The outputs are divided among the internal
demands and the external demands. This is stated as the balance
equation:

$$\mathbf{x} = A\mathbf{x} + \mathbf{d} \quad \text{(balance equation)}$$

Labor, energy (e.g., hydroelectric power, etc.), capital, and consumption are referred to as *below-the-line* items. They are determined
by the output vector:

$$L = \sum_{i=1}^{N} l_i x_i = \mathbf{l} \cdot \mathbf{x} \quad \text{(labor)}$$

$$E = \mathbf{e} \cdot \mathbf{x} \quad \text{(energy)}$$
$$K = \mathbf{k} \cdot \mathbf{x} \quad \text{(capital)}$$
$$C = \mathbf{c} \cdot \mathbf{x} \quad \text{(consumption)}$$

where the dot product notation is used. For example,

$$\mathbf{l} \cdot \mathbf{x} = \sum_{i=1}^{N} l_i x_i$$

where l_i denotes the number of labor units (e.g., worker-hours) needed to produce a unit of sector i; similarly for **e**, **k**, and **c**.

Prices are determined in the following way. Let p_i denote the price per unit of sector i, and let v_i denote the value added per unit of sector i. Value added is made up of labor, capital, and rents. The cost of producing a unit output of sector i, given the price vector **p**, is

$$\sum_{k=1}^{N} a_{ki} p_k$$

Thus, the cost of output **x** is $A^{Tr}\mathbf{p}$, where A^{Tr} denotes the transpose of the matrix A. The prices then satisfy the equation

$$\mathbf{p} = \mathbf{v} + A^{Tr}\,\mathbf{p}$$

These equations can be solved for **p** as

$$\mathbf{p} = (I - A^{Tr})^{-1}\,\mathbf{v}$$

if the matrix $(I - A^{Tr})$ is invertible. Thus, given the value added (**v**) and the technology (A), the price structure (**p**) can be found.

Also, given the technology (A) and the values added (**v**), this model shows what sector outputs and prices are needed to meet the external demand (**d**).

Input–output model of a multispecies fishery. We now apply these ideas to fishery management. Several species, each having several age classes, will be described at the beginning of the fishing period by the vector S_t. Fishing effort will be applied to these standing stocks (D_t) and the stocks that survive (Σ_t) will be available for spawning and, when appropriate, survival into the next year's standing stock. To fix ideas, we suppose that there are two species, say cod and haddock, each having three age classes. In addition, we will account for primary inputs to stocks that represent plankton and other energy sources utilized by the very young of these species. Schematically, the model is

	S_t	D_t	Σ_t		S_{t+1}
year t	↑	↑	↑	year $t+1$	↑
	standing stock	demand (fishing)	survivors for spawning		standing stock

The standing stock will be described by the vector

$$S = \begin{bmatrix} c_1 \\ c_2 \\ c_3 \\ h_1 \\ h_2 \\ h_3 \\ e_1 \\ e_2 \end{bmatrix}$$

where

c_1 = number of cod having age between one and two years

c_2 = number of cod having age between two and three years

c_3 = number of cod having age greater than three years

h_1 = number of haddock having age between one and two years

.

.

.

$e_1(e_2)$ = amount of primary energy available to cod (haddock) having age less than one year

The demand vector D_t accounts for landings (L_t), discards (Δ_t), foreign fishing (Φ_t) and other predators (P_t):

$$D_t = L_t + \Phi_t + \Delta_t + P_t$$

The standing stock in year $t(S_t)$ will be divided among fishing demand, natural mortality, and internal consumption by the other species. This gives the balance equation

$$S_t = AS_t + D_t + M\Sigma_t \quad \text{(balance equation)}$$

where A is the ecology matrix, which describes the prey–predator interactions between the species and age classes, and M determines the natural mortality of the standing stock throughout the year. Next year's ($t + 1$) standing stock will consist of survivors of the previous years' demand, predation, and natural mortality as well as new recruits that were laid as eggs in the past year ($t - 1$):

$$S_{t+1} = \Lambda\Sigma_t + R\Sigma_{t-1} \quad \text{(dynamic equation)}$$

The matrix Λ is the *life matrix*, which describes the survival probabilities of the various age classes and species, and the matrix R is the *renewal matrix*, which describes the fertility of spawners and the probability of survival of their offspring to age 1.

The ecology matrix has the form

$$A = \begin{pmatrix} A_c & A_{hc} & \\ A_{ch} & A_h & 0 \\ \hline \begin{matrix} e_{11} & 0 & 0 \\ 0 & 0 & 0 \end{matrix} & \begin{matrix} 0 & 0 & 0 \\ e_{22} & 0 & 0 \end{matrix} & \begin{matrix} 0 \\ 0 \end{matrix} \end{pmatrix}$$

where A_c is a 3×3 matrix describing the predation of cod on cod, A_{hc} is a 3×3 matrix describing the predation of haddock on cod ($A_{hc} = 0$), ..., e_{11} describes the number of units of e_1 needed to sustain a unit of yearling cod, and so on.

M is a diagonal mortality matrix with diagonal element $M_1, \ldots,$ $M_6, 0, 0$. M_1 gives the reciprocal of the probability that a yearling cod who survives demand survives to the spawning time, and so on.

Λ is the life matrix

$$\Lambda = \begin{bmatrix} \begin{matrix} 0 & 0 & 0 \\ \lambda^c_{12} & 0 & 0 \\ 0 & \lambda^c_{23} & \lambda^c_{33} \end{matrix} & & 0 \\ 0 & \begin{matrix} 0 & 0 & 0 \\ \lambda^h_{12} & 0 & 0 \\ 0 & \lambda^h_{23} & \lambda^h_{33} \end{matrix} & 0 \\ 0 & 0 & 0 \end{bmatrix}$$

where λ_{12} is the probability of surviving from the spawning period to the next year's standing stock of age 2, and so on.

The renewal matrix R has the form

$$R = \begin{bmatrix} \begin{matrix} 0 & 0 & b_c \\ 0 & 0 & 0 \\ 0 & 0 & 0 \end{matrix} & 0 & 0 \\ 0 & \begin{matrix} 0 & 0 & b_4 \\ 0 & 0 & 0 \\ 0 & 0 & 0 \end{matrix} & 0 \\ 0 & 0 & 0 \end{bmatrix}$$

where b_e gives the fertility of spawning adult cod multiplied by the probability of survival of eggs to age class 1, and so on.

Analysis of the model. The balance equation can be used to eliminate the spawning stock (Σ_t) from the system

$$\Sigma_t = M^{-1}(I - A)S_t - M^{-1}D_t$$

The inverse of M is defined on the species vector only. Thus, the dynamic equation becomes

$$S_{t+1} = \Lambda M^{-1}(I - A)S_t + R M^{-1}(I - A)S_{t-1}$$
$$- \Lambda M^{-1}D_t - R M^{-1}D_{t-1}$$

This equation can be solved by using the z-transform as described in Chapter 2.

Economics. The net revenue of the fishery is the price per unit landing times the number of units landed less the cost per unit effort times the effort used to obtain the landing. For stock j we have

$$R_j = p_j(L_j)L_j - cE_j$$

where a landing of size L_j produces an income $p_j(L_j)L_j$. The prices depend on the landing size, as shown in Figure 1.13. The more landed, the lower the price.

Using an effort–harvest relation such as that described in the preceding section, we have

Figure 1.13. Price per unit landing as a function of the number landed (L).

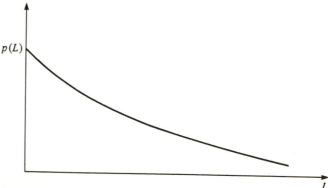

$$E_j = \frac{1}{q} \sum_{k=S_j-L_j-\Delta j-\phi j}^{S_j-\phi_j} \frac{1}{k}$$

(we ignore other predators). The question of discards is a difficult one. There are government limits on landings, so overcatch is discarded. Also, there are other species caught as by-catch, which are discarded to make room for the primary catch.

A more general input–output model of the economic aspects of a fishery can be formulated. The combined ecological–economic model can then be analyzed [see Hoppensteadt and Sohn (1980)].

Discussion. Fish biologists have data that can be used to estimate the coefficients in A, M, Λ, and R, as well as the effect of other predators on the standing stock. Parameter estimation is carried out using the least-squares method. Questions that can be asked include:

1 What effect will foreign fishing quotas and landing restrictions have on meeting the exogenous demands set by the economy?

2 What effect will various strategies for optimizing current revenue or discounted future revenue have on the fishery?

This model enables us to find answers to such questions in terms of what standing stocks are needed to meet demand requirements and what impact heavy harvesting of certain species will have on others.

E. *Summary*

Fishery management is quite complicated. First, fish biology is not well understood. In real fisheries there are many interacting species and age classes of fish. They are distributed in the sea, and their movements are complicated, in many cases completely unknown. Moreover, they live in various habitats and are exposed to a wide variety of nutrients and risks.

Second, the economics of a fishery are not well understood. Biological yield, current revenue, or discounted future revenues can be applied to evaluating the fishery strictly in only a few cases. In many instances, legal control is fragmented with contiguous jurisdictions having opposite laws. Social and political problems are often dominant, and these are difficult problems to grasp.

Still, the arguments summarized by the simple models in this section are quite powerful. They can provide some rough guidelines and suggest why certain management schemes lead to collapse of the fishery. This is clearly illustrated here when the difference is established between MSY, which gives a criterion for biological overfishing, and the population size that optimizes discounted future revenues, which gives a criterion for economic overfishing. This difference is profound if there is a reproduction curve having a predator pit, because sufficiently large discount rates can drive the fish population to extinction. In these cases, it is in the exploiter's best interest to drive the population to extinction through large harvests and then invest the proceeds.

Successively more elaborate models are being constructed and analyzed to explore many of these complications. As better data become available, more information will be gained. However, an important question is whether the knowledge gained can and will come in time to save many species now being exploited.

There is a growing literature on the management of fisheries. Introductions to this subject are given in books by Clark (1976) and Gulland (1974).

Appendix: least-squares approximation

Suppose that data are observed, say $(x_1, y_1), \ldots, (x_N, y_N)$, and that a theory predicts that these data lie on a straight line

$$y = ax + b$$

The parameters a and b must be chosen so as to best fit the data. The least-squares method gives one way of doing this.

First, we consider a measure of how far the data deviate from a straight line. Let

$$\Delta(a, b) = \sum_{i=1}^{N} (y_i - ax_i - b)^2$$

The parameters a and b are chosen to minimize this quantity. If a and b do give a minimum, then

$$\partial\Delta/\partial a = 0 \quad \text{and} \quad \partial\Delta/\partial b = 0$$

These two equations are

$$a \langle x^2 \rangle + b \langle x \rangle = \langle xy \rangle$$
$$a \langle x \rangle + bN = \langle y \rangle$$

where

$$\langle x^2 \rangle = \sum_{i=1}^{N} x_i^2, \quad \langle x \rangle = \sum_{i=1}^{N} x_i,$$

$$\langle y \rangle = \sum_{i=1}^{N} y_i, \quad \langle xy \rangle = \sum_{i=1}^{N} x_i y_i$$

The solutions of these equations are

$$a = \frac{N\langle xy \rangle - \langle x \rangle \langle y \rangle}{N\langle x^2 \rangle - \langle x \rangle^2} \quad \text{and} \quad b = \frac{\langle x^2 \rangle \langle y \rangle - \langle x \rangle \langle xy \rangle}{N\langle x^2 \rangle - \langle x \rangle^2}$$

These constants are then used in the straight-line formula to give the least-squares approximation.

Exercises

1.1 Swedish population data, 1780–1900, are given in the following table. Use the method of least squares to fit Malthus's model to these data. What is the growth rate? What is the predicted population size in 1920?

Data (thousands)

Year	1780	1800	1820	1840	1860	1880	1900
Population size	2104	2352	2573	3123	3824	4572	5117

1.2 The reproduction relation $P_{n+1} = 4P_n(1 - P_n)_+$ illustrates chaotic dynamics. This corresponds to $K\rho = 4$ in Section 1.1D.

(a) Make the substitution $P_n = \sin^2 \theta_n$ in this equation, and derive a reproduction curve for (θ_n, θ_{n+1}).

(b) Simulate the dynamics governed by the (θ_n) reproduction curve using the following method. Pick a value θ_1 at random from the interval $0 < \theta < \pi/2$. Evaluate the resulting sequence $\theta_1, \theta_2, \theta_3$, ..., θ_{100}. Count the number of times each of 100 equally spaced intervals covering $0 \leq \theta \leq \pi/2$ are hit by this sequence, and record these data. Repeat this calculation for 50 additional choices of θ_1. Plot the accumulation of the recorded data. A computer is recommended.

1.3 Consider a population that is described by the model $P_{n+1} = r_n P_n$, where r_n changes in each generation in a known way, say due to a changing environment. ($r_n \geq 0$.)

(a) Derive a formula for P_n in terms of P_1 and the sequence $r_1, \ldots, r_{n-1}$.

(b) Suppose that the sequence $\{r_n\}$ has period T; that is, there is an integer T such that $r_{n+T} = r_n$ for all n. Determine conditions under which $P_n \nearrow \infty$ and $P_n \searrow 0$ as $n \to \infty$.

(c) Suppose that the sequence $\{r_n\}$ has the property that

$$\lim_{N \to \infty} \frac{1}{N} \sum_{n=0}^{N-1} \log r_n = \log \bar{r}$$

exists. Show that if $\bar{r} < 1$, $P_n \searrow 0$, but if $\bar{r} > 1$, then $P_n \nearrow \infty$. Compare this with the result in part (b).

1.4 Consider a reproduction function f that has a static state P^*; that is, $f(P^*) = P^*$. Suppose that $|(df/dP)(P^*)| < 1$. Show that if P_0 is near P^*, the solution of $P_{n+1} = f(P_n)$, $n = 0, 1, 2, \ldots$, approaches P^* (i.e., $P_n \to P^*$ in succeeding generations).

1.5 Many of the models described here have continuous-time analogs. These are especially useful for considering populations that have overlapping generations. Let t denote a time variable chosen on a scale that is appropriate to the biology of the population being studied, and let $P(t)$ denote the population size at time t.

(a) The analog of Malthus's model is $dP/dt = \rho P$, where the constant ρ is the *intrinsic growth rate*. The constant ρ corresponds to $\log r$ in the discrete-time model. Solve this model and describe the dynamics.

(b) The Verhulst (or logistic) model is $dP/dt = \rho P(1 - P/K)$. K is called the carrying capacity of the population. Solve this model and describe the dynamics. [*Hint:* Let $P(t) = u(t)/v(t)$.]

(c) Consider Malthus's model with time-varying intrinsic growth rate, $dP/dt = \rho(t)P$. Evaluate $\lambda = \lim (t \to \infty) (1/t) \log P(t)$ in terms of ρ. Here λ, called the *Liapunov type number* for the equation, exists ($|\lambda| < \infty$) if ρ is periodic or almost periodic. Describe $P(t)$ if $\lambda = +\infty$ and $\lambda = -\infty$.

(d) Consider the model $dP/dt = \rho(t)P(1 - P/K)$. Suppose that $P(0) < K$, ρ has period T [i.e., $\rho(t + T) = \rho(t)$], and let $\bar{\rho} = (1/T) \int_0^T \rho(s) \, ds$. Show that

(1) If $\bar{\rho} < 0$, then $P(t) \to 0$.

(2) If $\bar{\rho} = 0$, then $P(t)$ is periodic, having period T.

(3) If $\bar{\rho} > 0$, then $P(t) \to K$.

1.6 Use the data in Table 1.2 and the general cicada model to simulate the evolution of the cicada population for $L = 7$ and $L = 13$. Use as initial data $P_n = 100$ for $n = 1, \ldots, L$.

2

Renewal theory and reproduction matrices

Census data are usually collected at five- or ten-year intervals, and these data give a profile of the population's age structure. There are several methods used to study age distribution and its effects on population changes, but the two basic approaches studied here are renewal theory and reproduction matrices. First, Fibonacci's sequence is used to illustrate the methods in an interesting way. After this, we develop the theories in the more general context of human populations.

The models described in this chapter are linear in their variables. They are extensions of the Malthus model to account for age structure. Obviously, they can only be expected to describe accurately a population's changes during certain growth phases. We have seen in Chapter 1 how limiting factors can be incorporated. However, these complications will not be described in this chapter. Instead, we focus on the widely used renewal and Leslie models.

2.1 Fibonacci sequence

Fibonacci (1202) described a hypothetical rabbit population that starts with a single pair, one male and one female. The pair will reproduce twice, and at each reproduction produce a new pair, one male and one female. These will go on to reproduce twice, and so on. Table 2.1 summarizes the dynamics.

A. *Fibonacci renewal equation*

Let B_n denote the number of newborn *females* at the nth reproduction time. Then for $n = 2, 3, 4, \ldots$,

$$B_{n+1} = B_n + B_{n-1} \qquad \text{(Fibonacci renewal equation)}$$

The initial conditions are $B_0 = 1$, $B_1 = 1$. Then $B_2 = 2$, $B_3 = 3$, B_4

31

= 5, and so on. Each term in the Fibonacci sequence $\{B_n\}$ is the sum of the two preceding terms.

This equation can be solved by setting $B_n = r^n$. It follows that r satisfies $r^2 - r - 1 = 0$, whose roots are

$$r_1 = \frac{1 + \sqrt{5}}{2} \quad \text{and} \quad r_2 = \frac{1 - \sqrt{5}}{2}$$

Since $B_0 = 1$ and $B_1 = 1$, we have that

$$B_n = \left(\frac{\sqrt{5} + 1}{2\sqrt{5}} \right) r_1^n + \left(\frac{\sqrt{5} - 1}{2\sqrt{5}} \right) r_2^n$$

This formula gives the number of females born at each reproduction time. Note that since $r_2/r_1 < 1$, we have that

$$\lim_{n \to \infty} r_1^{-n} B_n = \frac{\sqrt{5} + 1}{2\sqrt{5}} + \frac{\sqrt{5} - 1}{2\sqrt{5}} \lim_{n \to \infty} \left(\frac{r_2}{r_1} \right)^n$$

$$= \frac{\sqrt{5} + 1}{2\sqrt{5}}$$

We could say that

$$B_n \doteq r_1^n \left[\frac{\sqrt{5} + 1}{2\sqrt{5}} \right]$$

Table 2.1. *Number of female rabbits*

Time	Birth rate	Reproductive ages							
Age		0	1	2	3	4	5	6	7	8 ...
0	1								
1	1	1							
2	2	1	1						
3	3	2	1	1					
4	5	3	2	1	1				
5	8	5	3	2	1	1			
6	13	8	5	3	2	1	1		
7	21	13	8	5	3	2	1	1	
8	34	21	13	8	5	3	2	1	1

for large $n;$ the percent error made in this approximation is

$$\begin{array}{c}\text{percent}\\ \text{error}\end{array} = 100 \times \left| \frac{B_n - r_1^n\left(\dfrac{\sqrt{5}+1}{2\sqrt{5}}\right)}{B_n} \right|$$

$$= 100 \times \frac{1}{1 + \left(\dfrac{\sqrt{5}+1}{\sqrt{5}-1}\right)\left(\dfrac{r_1}{r_2}\right)^n} \longrightarrow 0 \qquad \text{as } n \to \infty$$

The numbers of reproducing pairs at any reproduction time can be easily determined directly from the sequence B_n. At the $(n + 1)$st reproduction time there will be B_n pairs reproducing for the first time and B_{n-1} for the second. A different point of view leads directly to the age structure (age is measured in units of reproduction intervals) through the reproduction ages.

B. *Fibonacci reproduction matrix*

Let $v_{0,n}$ = number of newborn females at the nth reproduction, and $v_{1,n}$ = number of females of age one at the nth reproduction. Only these two variables will be needed to illustrate the method. The notation is made concise by introducing the vector

$$\mathbf{v}_n = \begin{pmatrix} v_{0,n} \\ v_{1,n} \end{pmatrix}$$

Since

$$v_{0,n+1} = v_{0,n} + v_{1,n}$$
$$v_{1,n+1} = v_{0,n}$$

we define

$$M = \begin{pmatrix} 1 & 1 \\ 1 & 0 \end{pmatrix}$$

and simplify the notation by writing

$$\mathbf{v}_{n+1} = M\mathbf{v}_n \qquad \text{(Fibonacci reproduction model)}$$

The solution can be found by successive back substitutions

$$\mathbf{v}_{n+1} = M\mathbf{v}_n = M(M\mathbf{v}_{n-1}) = M^2\mathbf{v}_{n-1} = \ldots = M^n\mathbf{v}_1$$

where v_1 is the initial distribution vector

$$v_1 = \begin{pmatrix} 1 \\ 1 \end{pmatrix}$$

The age classes can therefore be described in terms of iterates of the matrix M (M^n) and the initial vector v_1. The eigenvalues of M play an important role in studying M^n. These are defined as the solutions of the equation

$$\det(M - \lambda I_2) = 0$$

where det is the determinant and I_2 is the 2×2 identity matrix

$$I_2 = \begin{pmatrix} 1 & 0 \\ 0 & 1 \end{pmatrix}$$

$$\det\begin{pmatrix} 1-\lambda & 1 \\ 1 & -\lambda \end{pmatrix} \equiv \lambda^2 - \lambda - 1 = 0$$

The solutions are

$$\lambda_1 = \frac{1 + \sqrt{5}}{2} \quad \text{and} \quad \lambda_2 = \frac{1 - \sqrt{5}}{2}$$

Note that $r_1 = \lambda_1$ and $r_2 = \lambda_2$.

The matrix M can be written in the form of its spectral decomposition

$$M = \lambda_1 P_1 + \lambda_2 P_2 \tag{1}$$

where

$$P_1 = \frac{1}{\lambda_1 + 2}\begin{pmatrix} \lambda_1^2 & \lambda_1 \\ \lambda_1 & 1 \end{pmatrix} \quad \text{and} \quad P_2 = \frac{1}{\lambda_2 + 2}\begin{pmatrix} \lambda_2^2 & \lambda_2 \\ \lambda_2 & 1 \end{pmatrix}$$

It follows that

$$P_1 P_1 = P_1, \quad P_2 P_2 = P_2, \quad \text{and} \quad P_1 P_2 = P_2 P_1 = Z_2$$

where Z_2 is the 2×2 zero matrix

$$Z_2 = \begin{pmatrix} 0 & 0 \\ 0 & 0 \end{pmatrix}$$

These formulas are easy to verify. For example,

$$P_1P_1 = \frac{1}{(\lambda_1 + 2)^2} \begin{pmatrix} \lambda_1^4 + \lambda_1^2 & \lambda_1^3 + \lambda_1 \\ \lambda_1^3 + \lambda_1 & \lambda_1^2 + 1 \end{pmatrix}$$

but

$$\frac{\lambda_1^4 + \lambda_1^2}{(\lambda_1 + 2)^2} = \frac{\lambda_1^2(\lambda_1^2 + 1)}{(\lambda_1 + 2)^2} = \frac{\lambda_1^2(\lambda_1 + 2)}{(\lambda_1 + 2)^2} = \frac{\lambda_1^2}{\lambda_1 + 2}$$

since $\lambda_1^2 + 1 = \lambda_1 + 1 + 1 = \lambda_1 + 2$. The other components are verified in the same way. To see that $P_1P_2 = Z_2$, we first evaluate the product

$$P_1P_2 = \frac{1}{(\lambda_1 + 2)(\lambda_2 + 2)} \begin{pmatrix} \lambda_1^2\lambda_2^2 + \lambda_1\lambda_2 & \lambda_1^2\lambda_2 + \lambda_1 \\ \lambda_1\lambda_2^2 + \lambda_2 & \lambda_1\lambda_2 + 1 \end{pmatrix}$$

Since

$$\lambda_1\lambda_2 = \frac{1 + \sqrt{5}}{2}\frac{1 - \sqrt{5}}{2} = \frac{1 - 5}{4} = -1$$

all the components in this matrix are zero.

Matrices P_1 and P_2 satisfying these relations are called *projection matrices*. Note that

$$M^2 = (\lambda_1P_1 + \lambda_2P_2)(\lambda_1P_1 + \lambda_2P_2)$$
$$= \lambda_1^2P_1^2 + \lambda_1\lambda_2P_1P_2 + \lambda_2\lambda_1P_2P_1 + \lambda_2^2P_2^2$$
$$= \lambda_1^2P_1 + \lambda_2^2P_2$$

It follows that for any integer n,

$$M^n = \lambda_1^nP_1 + \lambda_2^nP_2$$

This formula converts the problem of evaluating M^n, a matrix multiplication problem, to evaluating λ_1^n and λ_2^n, two scalar problems.

The age distribution vector is given by

$$v_{n+1} = M^nv_1 = \lambda_1^nP_1v_1 + \lambda_2^nP_2v_1$$

which gives an easily calculated formula for v_n.

One final calculation will be performed on this example. The preceding formula shows that

$$\lambda_1^{-n} \mathbf{v}_{n+1} = P_1 \mathbf{v}_1 + \left(\frac{\lambda_2}{\lambda_1}\right)^n P_2 \mathbf{v}_2$$

Since $\lambda_2/\lambda_1 = (\sqrt{5} - 1)/(\sqrt{5} + 1) < 1$, we see that

$$\lim_{n \to \infty} \lambda_1^{-n} \mathbf{v}_{n+1} \doteq P_1 \mathbf{v}_1$$

The vector

$$P_1 \mathbf{v}_1 = \frac{1}{\lambda_1 + 2} \begin{pmatrix} \lambda_1^2 v_{1,1} + \lambda_1 v_{2,1} \\ \lambda_1 v_{1,1} + v_{2,1} \end{pmatrix} = \frac{1}{\lambda_1 + 2} \begin{pmatrix} 2\lambda_1 + 1 \\ \lambda_1 + 1 \end{pmatrix}$$

is called the *stable age distribution* of the population. This because for large n

$$\mathbf{v}_{n+1} \doteq \lambda_1^n P_1 \mathbf{v}_1$$

where the percent error in this approximation is neglibible. Therefore, the ratio

$$v_{0,n} : v_{1,n} \doteq 2\lambda_1 + 1 : \lambda_1 + 1 = 1.618$$

which is independent of n. Thus, the ratio of the age classes remains constant even though the numbers in these classes grow geometrically. Let us check this for the Fibonacci sequence. Since $\lambda_1 = (1 + \sqrt{5})/2$,

$$(2\lambda_1 + 1)/(\lambda_1 + 1) \doteq 1.618$$

Table 2.2 shows the Fibonacci sequence and ratio of contiguous terms. This shows that for $n \geq 9$, $B_{n+1} \doteq 1.618\, B_n$. Therefore, the Fibonacci sequences, although a sequence of integers, is approximately a geometric sequence with common ratio $= 1.618$. From the renewal equation, we know that $B_{n+1}/B_n \doteq r_1$.

Table 2.2. *Fibonacci sequence and ratio contiguous terms*

Sequence (B_n)	1	1	2	3	5	8	13	21	34	55	89
Ratio (B_{n+1}/B_n)		1	2	1.5	1.67	1.625	1.615	1.619	1.618	1.618	1.618	1.618

2.2   Renewal equation

Human population data collected at five-year intervals give a dynamic profile of age structure. This can be described in terms of the birth rate or in terms of the age distribution vector. We begin by studying the birth rate. Let B_n = number of female births in the nth census period. The proportion of the births in the nth census that survive to the next census will be denoted by σ_1, the proportion surviving from that group to the next census is σ_2, and so on. The female population sizes are given in Table 2.3.

The fertility of those in the mth age class will be denoted by b_m. For example, the number of female births in the $(n + 5)$th census due to those born in the nth census will be

$$b_5\sigma_5\sigma_4\sigma_3\sigma_2\sigma_1 B_n$$

Finally, let

$$l_k = \sigma_k\sigma_{k-1} \ldots \sigma_1$$

denote the probability of surviving from birth to the kth age class. Census reports tabulate l, the result is called the *life table*.

The population age structure in the first census will be denoted by

$$\mathbf{u} = (u_0, u_1, u_2, \ldots)$$

Table 2.4, which shows the evolution of this distribution, indicates how the population age classes change initially. We can summarize the process in a single equation:

$$B_n = f_n + \sum_{k=1}^{n} b_k l_k B_{n-k} \qquad \text{(renewal equation)}$$

where f_n is the contribution to births in the nth census by the initial population that has survived to the nth census.

$$f_n = \sum_{j=n+1}^{\infty} b_j l_j u_{j-n}/l_{j-n}$$

A. *Analysis*

It is useful to transform the renewal equation by introducing a generating function. Let

Table 2.3. *Female population sizes by census*

Census \ Age class	0 $0 \leq a < 5$	1st $5 \leq a < 10$	2nd $10 \leq a < 15$	3rd $15 \leq a < 20$	4th $20 \leq a < 25$	5th $25 \leq a < 30$	etc.
n	B_n						
$n+1$		$\sigma_1 B_n$					
$n+2$			$\sigma_2\sigma_1 B_n$				
$n+3$				$\sigma_3\sigma_2\sigma_1 B_n$			
$n+4$					$\sigma_4\sigma_3\sigma_2\sigma_1 B_n$		
$n+5$						$\sigma_5\sigma_4\sigma_3\sigma_2\sigma_1 B_n$	
.							
.							
.							

$$\tilde{B}(s) = \sum_{k=0}^{\infty} B_k s^k, \quad \tilde{f}(s) = \sum_{k=0}^{\infty} f_k s^k, \quad \tilde{p}(s) = \sum_{k=0}^{\infty} b_k l_k s^k$$

Since there is a largest age of reproduction, $b_k = 0$ for $k > A$, where A is the last age class of fertility. Therefore, the sums in $\tilde{f}$ and $\tilde{p}$ are (finite) polynomials in s. We assume that $B_0 = f_0 = u_0$ and that $b_0 = 0$.

Multiplying both sides of the renewal equation by s^n and summing for $n = 0$ to ∞, we obtain the transformed equation

$$\tilde{B}(s) = \tilde{f}(s) + \tilde{p}(s)\tilde{B}(s)$$

This can be solved for $\tilde{B}$ as

$$\tilde{B}(s) = \tilde{f}(s)/(1 - \tilde{p}(s))$$

This defines a smooth function of s as long as $\tilde{p}(s) \neq 1$. There are A roots for this, say $s_1, \ldots, s_A$.

We first suppose that the roots are distinct. B_n can be recovered from the generating function by the inverse Laplace transform formula,

$$B_n = \frac{1}{2\pi i} \oint \frac{s^{-n-1}\tilde{f}(s)}{1 - \tilde{p}(s)} \, ds$$

where the contour of integration is chosen as any curve enclosing the zeros of the denominator in the complex plane. Thus,

$$B_n = \sum_{a=1}^{A} c_a s_a^{-n-1}$$

where

$$c_a = (\tilde{f}(s_a)/\tilde{p}'(s_a))$$

Table 2.4. *Evolution of the population age structure*

Census \ Age class	0	1st	2nd	3rd	4th	5th
0	u_0	u_1	u_2	u_3	u_4	u_5
1	B_1	$\sigma_1 u_0$	$\sigma_2 u_1$	$\sigma_3 u_2$	$\sigma_4 u_3$	$\sigma_5 u_4$
2	B_2	$\sigma_1 B_1$	$\sigma_2 \sigma_1 u_0$	$\sigma_3 \sigma_2 u_1$	$\sigma_4 \sigma_3 u_2$	$\sigma_5 \sigma_4 u_3$

An interesting derivation of this result is given by Charlesworth (1980).

The general case is described by Feller (1968), and it is summarized in the renewal theorem. Let

$$p = \sum_{k=1}^{A} b_k l_k, \qquad f = \sum_{k=1}^{A} f_k$$

Then we have:

Renewal theorem (Feller, 1968)

(a) If $p < 1$, then $B_n \to 0$ and $\sum_{k=0}^{\infty} B_k = f/(1 - p)$.

(b) If $p = 1$, then $B_n \to f/\tilde{p}'(1)$.

(c) If $p > 1$ and if $\tilde{p}(s) - 1$ has a unique positive root ρ^* with all other roots satisfying $|\rho| > \rho^*$ then

$$\rho^{*k} B_k \to \tilde{f}(\rho^*)/(\rho^* p'(\rho^*))$$

If the population is viable, then it must be that $p > 1$. Also, $\tilde{p}(s) = 1$ is called the *characteristic equation* and the root ρ^* is the dominant characteristic root. When $p > 1$, the theorem shows that B_n increases roughly at the rate ρ^{*-1}. Note that $\rho^* < 1$ because $\tilde{p}(1) > 1$ and $\tilde{p}(0) < 1$. The term $1/\rho^*$ is called the *intrinsic growth rate*. A variant of the characteristic equation was derived by Euler (1760).

The generating functions are the analog of the Laplace transform for discrete equations. They are sometimes referred to as z-transforms, but then z^{-1} is used in place of s.

2.3   Reproduction matrix: honest matrices

The age profile at the nth census is now described by the numbers $v_{0,n}, v_{1,n}, \ldots, v_{A-1,n}$. We ignore the age classes beyond the reproductive ones. This information is summarized in the vector

$$\mathbf{v}_n = \begin{bmatrix} v_{0,n} \\ \cdot \\ \cdot \\ \cdot \\ v_{A-1,n} \end{bmatrix}$$

The age profile at the next census is determined from the equations

$$v_{0,n+1} = \sigma_1 b_1 v_{0,n} + \sigma_2 b_2 v_{1,n} + \ldots + \sigma_A b_A v_{A-1,n}$$
$$v_{1,n+1} = \sigma_1 v_{0,n}$$
$$v_{2,n+1} = \sigma_2 v_{1,n}$$
$$v_{A-1,n+1} = \sigma_{A-1} v_{A-1,n}$$

These equations can be made more concise by using vector notation,

$$\mathbf{v}_{n+1} = M\mathbf{v}_n$$

where the matrix M, called the *reproduction matrix*, has the form

$$M = \begin{bmatrix} \sigma_1 b_1 & \sigma_2 b_2 & \ldots & \sigma_A b_A \\ \sigma_1 & 0 & \ldots & 0 \\ 0 & \sigma_2 & \ldots & 0 \\ \cdot & \cdot & & \cdot \\ \cdot & \cdot & & \cdot \\ \cdot & \cdot & & \cdot \\ 0 & 0 & \sigma_{A-1} & 0 \end{bmatrix}$$

The profile in the nth census is therefore given by

$$\mathbf{v}_n = M^n \mathbf{v}_0 \qquad \text{(Leslie model)}$$

It remains to unravel the matrix M^n. This is done by using the spectral decomposition of M, and we make use of the Perron–Froebenius theory to do this.

First, suppose that we know the A eigenvalues of M and that they are distinct, say $\lambda_1, \ldots, \lambda_A$. Then there are A column vectors $\phi_1, \ldots, \phi_A$ such that

$$M\phi_i = \lambda_i \phi_i \quad \text{for} \quad i = 1, \ldots, A$$

ϕ_1 is the right eigenvector associated with λ_1, and so on. The eigenvectors $\phi_1, \ldots, \phi_A$ span the space in the sense that given a vector $\mathbf{v}_0$, there are constants $c_1, \ldots, c_A$ such that

$$\mathbf{v}_0 = c_1 \phi_1 + \ldots + c_A \phi_A$$

Applying M to $\mathbf{v}_0$ gives

$$M\mathbf{v}_0 = M(c_1 \phi_1 + \ldots + c_A \phi_A) = c_1 \lambda_1 \phi_1 + \ldots + c_A \lambda_A \phi_A$$

and

$$M^2 \mathbf{v}_0 = M(M\mathbf{v}_0) = c_1 \lambda_1^2 \phi_1 + \ldots + c_A \lambda_A^2 \phi_A$$

and so on. In general,

$$M^n \mathbf{v}_0 = c_1 \lambda_1^n \phi_1 + \ldots + c_A \lambda_A^n \phi_A$$

This is a form of the spectral decomposition of M^n. Suppose the eigen-values are numbered so that $|\lambda_1| > |\lambda_2| > \ldots > |\lambda_A|$. Then $\lambda_1^{-n} M^n \mathbf{v}_0 \to c_1 \phi_1$.

In the general case, where the eigenvalues $\lambda_1, \ldots, \lambda_A$ are not nec-essarily distinct, the Perron–Froebenius theorem gives useful information.

A technical condition is needed to state this result. We say that the matrix M is *honest* if the characteristic polynomial

$$\det(M - \lambda I) = \sum_{k=0}^{A} \alpha_k \lambda^k$$

has the property that there are at least two nonzero coefficients, say $\alpha_{k'}$ and $\alpha_{k''}$, with k' and k'' relatively prime (i.e., k' and k'' have no common integer factors other than 1). For example, the polynomial $\lambda^4 + \lambda$ is honest (α_1 and $\alpha_4 \neq 0$) but $\lambda^8 + \lambda^4$ is not (8 and 4 have 4 as a common factor).

Perron–Froebenius theorem. If an $A \times A$ matrix Q is an honest matrix having all components nonnegative, then there is a positive real eigenvalue λ^*. In addition, all other eigenvalues λ satisfy $|\lambda| < \lambda^*$. Finally, the matrix Q^n can be written as a sum

$$Q^n = \lambda^{*n} P + E_n$$

where $|E_n|/\lambda^{*n} \to 0$ as $n \to \infty$ and the matrix P takes vectors into their projection onto the (right) eigenvector corresponding to λ^*.

Applying this result to M, we see that if M is honest there is a dominant eigenvalue λ^* and

$$M^n \mathbf{v}_0 = c_* \lambda^{*n} \phi_* + E_n$$

It follows that

$$\lambda^{*-n} M^n \mathbf{v}_0 \to c_* \phi_* \quad \text{as} \quad n \to \infty$$

These calculations show that the eigenvalues of the reproduction matrix play a critical role in describing the population's age structure after many generations. In particular,

$$M^n v_0 \doteq \lambda^{*n} c_* \phi_*$$

It follows that the components of the vector $M^n v_0$ eventually have ratios that are independent of n:

$$(M^n v_0)_1 : (M^n v_0)_2 : \ldots : (M^n v_0)_A = c_* \phi_{1*} : c_* \phi_{2*} : \ldots : c_* \phi_{A*}$$

Thus, the distribution of the population among the age classes remains constant even as the population grows (if $\lambda^* > 1$) or dies out ($\lambda^* < 1$). For this reason the vector ϕ^* is called the *stable age distribution.*

The reproduction model $v_{n+1} = M v_n$ is quite similar in spirit to the Malthus model. The largest real eigenvalue corresponds to the intrinsic growth rate. This model has been derived in various forms by Leslie (1945, 1948) Bernardelli (1942), McKendrick (1926), Lotka (1922), and Euler (1760).

2.4 Population waves: dishonest matrices

Bernardelli (1942) pointed out that if fertility is concentrated in one age group, then the birth rate could be a periodic function of time. This is an extreme example of a population wave. The birth rate will appear to oscillate if several characteristic roots are imaginary but have the same modulus as the dominant real root. This can occur if M is a dishonest matrix. The following example illustrates this phenomenon.

We take $A = 8$, but $b_i = 0$ for $i \neq 4, 8$. The reproduction matrix is

$$M = \begin{bmatrix} 0 & 0 & 0 & b_4 & 0 & 0 & 0 & b_8 \\ \sigma_1 & 0 & 0 & 0 & 0 & 0 & 0 & 0 \\ 0 & \sigma_2 & 0 & 0 & 0 & 0 & 0 & 0 \\ 0 & 0 & \sigma_3 & 0 & 0 & 0 & 0 & 0 \\ 0 & 0 & 0 & \sigma_4 & 0 & 0 & 0 & 0 \\ 0 & 0 & 0 & 0 & \sigma_5 & 0 & 0 & 0 \\ 0 & 0 & 0 & 0 & 0 & \sigma_6 & 0 & 0 \\ 0 & 0 & 0 & 0 & 0 & 0 & \sigma_7 & 0 \end{bmatrix}$$

The eigenvalues of this matrix are easily calculated. The determinant

$\det(M - \lambda I_8)$ can be expanded by minors to give

$$\lambda^8 - \alpha_4\lambda^4 - \alpha_8 = 0$$

where $\alpha_4 = b_4\sigma_1\sigma_2\sigma_3$ and $\alpha_8 = b_8\sigma_1 \cdots \sigma_7$. This equation is a quadratic in λ^4 and the solutions are

$$\lambda_+^4 = \frac{\alpha_4 + (\alpha_4^2 + 4\alpha_8)^{1/2}}{2} = \rho^4$$

$$\lambda_-^4 = \frac{\alpha_4 - (\alpha_4^2 + 4\alpha_8)^{1/2}}{2} = (\sqrt{i\mu})^4$$

Thus, $\lambda_+ = \rho,\ \rho\omega_4,\ \rho\omega_4^2,\ \rho\omega_4^3,\ \lambda_- = \mu e^{i\pi/4},\ \mu e^{i\pi/4}\omega_4,\ \mu e^{i\pi/4}\omega_4^2$, and $\mu e^{i\pi/4}\omega_4^3$, where ω_4 is the generator of the fourth roots of unity ($\omega_4 = e^{i\pi/2}$). For this example, if $\rho = 1$, then four of these roots have magnitude 1.

When $\rho = 1$, the solution of this example is *periodic;* this is an extreme form of a *Bernardelli population wave* (i.e., a periodic age structure). For $\rho > 1$ or $\rho < 1$, but ρ near 1, the solution grows or decays, but in an oscillatory way. This behavior is also referred to as a population wave. Population waves are observed in human populations [see Keyfitz and Flieger (1971)].

Exercises

2.1 *McKendrick's model.* Let $u(a, t)$ denote the age distribution of a population at time t. The total population is given by

$$\int_0^\infty u(a, t)\ da$$

In a time step of size h, the change in the age distribution is assumed to be given by

$$u(a + h, t + h) - u(a, t) = -d(a, t)hu(a, t)$$

where $d(a, t)$ is the death rate of individuals of age a at time t.

(a) Show that if u is a smooth function, then

$$\frac{\partial u}{\partial t} + \frac{\partial u}{\partial a} = -d(a, t)u$$

The birth rate is given by $u(0, t)$, and we suppose it to be related to the age distribution by

$$u(0, t) = \int_0^\infty \beta(a, t)u(a, t)\ da$$

where $\beta(a, t)$ is the (crude) birth rate of individuals of age a at time t. The initial age distribution

$$u(a, 0) = \mathring{u}(a)$$

is given.

(b) Suppose that $u(0, t) = B(t)$ is given. Solve the problem

$$\frac{\partial u}{\partial t} + \frac{\partial u}{\partial a} = -d(a, t)u, \qquad u(0, t) = B(t), \quad u(a, 0) = \mathring{u}(a)$$

(c) Use the solution in part (b) to reduce the problem in part (a) to a single equation for the birth rate $B(t)$. Put this in the form

$$B(t) = f(t) + \int_0^t k(a, t)B(t - a) \, da$$

by identifying f and the kernel k.

If $\beta(a, t) = 0$ for $a \geq A$ and all t, show $f(t) \to 0$ as $t \to \infty$.

(d) Suppose that d and β depend only on a, not t. Then show that k depends only on a. Thus,

$$B(t) = f(t) + \int_0^t k(a)B(t - a) \, da \qquad \text{(renewal equation)}$$

Define the Laplace transform of a function g by

$$\tilde{g}(s) = \int_0^\infty e^{-st} g(t) \, dt$$

Show that

$$\tilde{B}(s) = \tilde{f}(s)/(1 - \tilde{k}(s))$$

(e) Show that $1 - \tilde{k}(s)$ has a unique real solution, say s_0, and that all other roots have real part less than s_0.

(f) The inverse Laplace transform is defined by

$$g(t) = (1/2\pi i) \int_{c-i\infty}^{c+i\infty} e^{st} \tilde{g}(s) \, ds$$

where the contour of integration is chosen to the right of all singularities of $\tilde{g}(s)$. Suppose that $\tilde{k}(s) = 1$ has only simple roots. Show that

$$B(t) = \sum_{m=0}^\infty A_m \exp(s_m t)$$

where s_m are solutions of the characteristic equation

$$1 = \int_0^\infty e^{-sa} k(a) \, da \qquad [= \tilde{k}(s)]$$

(g) Show that $\lim (t \to \infty) \exp(-s_0 t) u(a, t)$ exists for each a. This limit is referred to as the population's *stable age distribution*.

2.2 Show that the stable age vector ϕ^* of the Leslie model is given by

$$
\phi_* = \begin{bmatrix} (\lambda^*)^{A-1}/\sigma_1 \ldots \sigma_{A-1} \\ (\lambda^*)^{A-2}/\sigma_2 \ldots \sigma_{A-1} \\ . \\ . \\ . \\ \lambda^*/\sigma_{A-1} \\ 1 \end{bmatrix}
$$

where λ^* is the dominant eigenvalue.

2.3 Let P_n denote the number of adults of reproductive age in a population in which newborns take k time periods to mature. This population is described by the nonlinear renewal equation

$$ P_{n+1} = \sigma P_n + f(P_{n-k}) $$

where σ is the survival probability of adults and f describes reproduction and probability of surviving k periods. Let P^* be a static state

$$ P^* = \sigma P^* + f(P^*) $$

(a) Show that if all roots of the characteristic equation

$$ s^{k+1} = \sigma s^k + f'(P^*) $$

lie inside the unit circle, then P^* is stable; that is, if P_0 is near P^*, then $P_n \to P^*$ as $n \to \infty$.

(b)

Show that if $|f'(P^*)| < 1 - \sigma$, then P^* is stable.

3

Markov chains

3.1   Bacterial genetics

Bacteria are single-celled organisms that are enclosed within the cell wall; the interior is made up of cytoplasmic material and it contains the various mechanisms needed for cell life and reproduction. In particular, there is the *chromosome*, a circular loop of deoxyribonucleic acid (DNA), which carries a code for all cell functions. The chromosome is a double helix like a twisted ladder, the rungs correspond to pairs of nucleic acids: that is, pairs of the bases adenine, thymine, cytosine, and guanine. The only possible pairs are AT and GC. Therefore, the sequence of base pairs can be labeled by one strand. The complement strand is made by replacing each A by T, each G by C, and so on.

A *gene* is a segment of the chromosome that codes for production by the cell of a product, such as an enzyme. The location of a gene on the chromosome is called its *locus*, and variant forms of the gene are called *alleles*.

Many types of bacteria have additional genetic material called extrachromosomal elements or *plasmids*. Plasmids are small circular pieces of DNA that also carry genes. However, plasmids can pass from cell to cell, and some genes can "jump" from plasmids to chromosomes.

The cell cycle begins with a newborn daughter. All of the cell's components, including the chromosome, are *replicated*. The replication is followed by splitting or division of the cell into two daughters, each receiving one replicate.

First, we will model frameshift mutations on the chromosome, and then the replication and distribution of plasmids among daughters.

A. *Frameshift mutation: a two-state Markov chain*

A segment of DNA is processed in several ways during the cell cycle: It is replicated during reproduction, it is copied during transcription, and so on. Errors can occur in these processes. For example, during replication a nucleotide can be deleted or inserted. This can have a dramatic effect on the message carried by the chromosome. Table 3.1 illustrates one possibility. The nucleotides are read in triples; each triple codes for an amino acid. Deletion of one nucleotide from the sequence results in an entirely new sequence past the point of deletion, in this case one that has a transcription termination triple appearing prematurely (UAG). Such a change is called a *frameshift mutation.*

Let ϵ = probability of a frameshift replication error at a given nucleotide during replication of the chromosome, and ρ = probability of repair—that a correct nucleotide is inserted immediately after a given nucleotide or that a given incorrect nucleotide is deleted during replication. Suppose that a gene is made up of a number N of nucleotides. Then the probability that a frameshift error occurs in the gene is ϵN = probability of a frameshift error occurring in the gene. The probability that such an error is repaired is ρ = probability of repair of a frameshift error. We write $\mu = \epsilon N$ = mutation probability for a given gene, and $\nu = \rho$ = reverse mutation probability.

A population of bacteria can be divided into two groups, mutants, and wild types, depending on whether they have or have not suffered a frameshift mutation in a given gene. An example of this is the bacterium *Salmonella typhimurium,* which has genes for the production of the amino acid histidine. The wild types are able to produce histidine and are labeled His$^+$; the mutants are unable to produce histidine and are labeled His$^-$.

Let $p_{W,n}$ = proportion of the nth generation of wild type, and $p_{M,n}$ = proportion of the nth generation of mutant type. With this notation we can write

$$p_{W,n+1} = (1 - \mu)p_{W,n} + \nu p_{M,n}$$
$$p_{M,n+1} = \mu p_{W,n} + (1 - \nu)p_{M,n}$$

or writing

$$\mathbf{p}_n = (p_{W,n}, p_{M,n}) \quad \text{and} \quad P = \begin{bmatrix} 1 - \mu & \mu \\ \nu & 1 - \nu \end{bmatrix}$$

we have

$$\mathbf{p}_{n+1} = \mathbf{p}_n P \quad \text{(two state chain)} \tag{1}$$

Table 3.1. *Example of a replication error*

	ACG	AAA	CCG	AAG	CAT	CTT	A . . .
DNA sequence	AAA						
RNA sequence[a]	UUU	UGC	UUU	GGC	UUC	GUA	GAA
Amino acid sequence	Phenyl-alanine	Cysteine	Phenyl-alanine	Glycine	Phenyl-alanine	Valine	Glutamic acid
Deletion		↑					
New DNA sequence	AAA	CGA	AAC	CGA	AGC	ATC	TTA
New amino acid sequence	Phenyl-alanine	Alanine	Leucine	Alanine	Serine	Stop	

[a]U denotes uracil, which replaces thymine in RNA.

The matrix P is called the *transition probability matrix* and the vector $\mathbf{p}_n$ is the nth generation's *probability distribution vector: $p_{W,n}$* gives the probability that a given cell selected from the nth generation will be of the wild type. The row vector format is used for notational convenience; for example, if P_{ij} denotes the element of P in the ith row and jth column, then P_{ij} gives the probability of passing from state i (W or M, in this example) to state j (again W or M) in one generation. P is also a *stochastic matrix,* meaning that it has nonnegative entries and that the row sums are equal to 1 ($\sum_{j=1}^{2} P_{ij} = 1$).

The sequence x_n = cell type of a random sample from the nth generation is a *Markov chain.* It has two possible states W or M, so it is called a two-state Markov chain. Interesting properties of the Markov chain can be determined from the transition probability matrix. Equation (1) can be solved by successive back substitutions

$$\mathbf{p}_n = \mathbf{p}_{n-1} P^{n-1} = \ldots = \mathbf{p}_0 P^n$$

As in Chapter 2, if we denote the eigenvalues of P by λ_1, λ_2, the eigenvalues are

$$\lambda_1 = 1, \qquad \lambda_2 = 1 - (\mu + \nu)$$

In practice, $0 < \mu, \nu < 1$, so $|\lambda_2| < 1$. Since $\lambda_1 \neq \lambda_2$, there are constants c_1 and c_2 such that

$$\mathbf{p}_0 P = c_1 \lambda_1 \phi_1 + c_2 \lambda_2 \phi_2$$

where the vectors ϕ_1 and ϕ_2 are the (left) eigenvectors corresponding to λ_1 and λ_2, respectively. This formula is the spectral decomposition of P. Therefore,

$$\mathbf{p}_n = \mathbf{p}_0 P^n = c_1 \lambda_1^n \phi_1 + c_2 \lambda_2^n \phi_2$$

Dividing by λ_1^n and passing to the limit $n = \infty$, we have

$$\lim_{n \to \infty} \lambda_1^{-n} \mathbf{p}_n = c_1 \phi_1$$

Now, the left eigenvector corresponding to $\lambda_1 = 1$ is $\phi_1 = (\mu/\nu, 1)$. Therefore,

$$c_1 \phi_1 = (p_{W,0} + p_{M,0})(\mu/\nu, 1) = (\mu/\nu, 1)$$

In particular, the ratio of wild types to mutants will be ν/μ after many generations. Thus, the eigenvector corresponding to the eigenvalue $\lambda = 1$ gives the asymptotic distribution of the population.

B. *Plasmid incompatibility: hypergeometric and Polya chains*

Regular replication, regular partitioning. We consider a plasmid P and two homogenic derivatives P' and P''. The population distribution of the classes of cells with various $P':P''$ ratios in any generation can be calculated from the distribution in the preceding generation. The plasmid P is assumed to appear with copy number N in all newborn cells and to replicate according to an

$$N \xrightarrow{\text{replication}} 2N \xrightarrow{\text{partitioning}} N$$

cycle. We suppose for simplicity that the division cycle of all cells is synchronized.

Let p_i = proportion of newborn cells having i P'-plasmids and $N - i$ P''-plasmids. After replication, there will be $2N$ plasmids, $2i$ of them P'- and $2N - 2i$ of them P''-plasmids.

There are a number of ways that these $2N$ plasmids can be distributed among the two daughters. We suppose that each daughter receives N copies. j copies of P' can be selected for one daughter in $\begin{pmatrix} 2i \\ j \end{pmatrix}$ different ways. This notation is the usual binomial coefficient, where

$$\begin{pmatrix} m \\ n \end{pmatrix} = \frac{m!}{n!(m-n)!} \quad \text{if} \quad 0 \le n \le m$$
$$= 0 \qquad \text{for other } m, n$$

The total number of partitions is $\begin{pmatrix} 2N \\ N \end{pmatrix}$, and the number of ways one daughter can have j P'- and $N - j$ P''-plasmids is

$$\begin{pmatrix} 2i \\ j \end{pmatrix} \begin{pmatrix} 2N - 2i \\ N - j \end{pmatrix}$$

if the mother originally had i P'- and $N - i$ P''-plasmids. Therefore,

$$\pi_{ij} = \frac{\begin{pmatrix} 2i \\ j \end{pmatrix}\begin{pmatrix} 2N - 2i \\ N - j \end{pmatrix}}{\begin{pmatrix} 2N \\ N \end{pmatrix}} \qquad \text{(hypergeometric chain)} \quad (2)$$

gives the probability of one daughter having j P'- and $N - j$ P''-plasmids. Since the reverse combination is also possible, the probability of an i P'-plasmid mother bearing one daughter with j and one with $N - j$ P'-plasmids is $2\pi_{ij}$.

If there are M bacteria in the nth generation, then after replication and division there will be $2M$ of them. If $p_{j,n+1}$ denotes the proportion having j P'' cells among the daughter [$(n + 1)$st] generation, then

$$2Mp_{j,n+1} = \sum_{i=0}^{N} (Mp_{i,n})(2\pi_{ij})$$

Thus,

$$p_{j,n+1} = \sum_{i=0}^{N} p_{i,n}\pi_{ij} \quad \text{for} \quad j = 0, \ldots, N$$

If we set

$$\mathbf{p}_n = (p_{0,n}, p_{1,n}, \ldots, p_{N,n}) \quad \text{and} \quad \Pi = (\pi_{ij})_{i,j=0}^{N}$$

then $\mathbf{p}_{n+1} = \mathbf{p}_n\Pi$. The transition probability matrix Π is made up of hypergeometric probabilities for sampling without replacement. It has been widely used in other areas of genetics that are formally similar to plasmid systems [see Feller (1968)].

Selective growth. A cell having at least one copy of each plasmid is referred to as a *heteroplasmid cell*. The plasmids might govern resistance to antibiotics, say P' gives resistence to streptomycin and P'' resistance to ampicillin. A typical segregation experiment involves introduction of a single P' plasmid into a cell carrying several copies of another plasmid P''. The clone is first grown on a medium that selects for both plasmids (say, containing both streptomycin and ampicillin) and then transferred to a nonselective growth medium with scoring for the number of heteroplasmid cells in each generation.

We first describe cell growth under selective conditions. Starting with a proportion p_i of newborn cells having i P'-plasmids, the proportion of the population after division having j P'-plasmids now is

$$p_{j,n+1} = \frac{\text{number of daughters with } j \ P'\text{-plasmids}}{\text{number of heteroplasmid cells in daughter population}}$$

$$= \frac{\sum\limits_{i=1}^{N-1} p_{i,k}\pi_{ij}}{\sum\limits_{k=1}^{N-1}\sum\limits_{m=1}^{N-1} p_{k,n}\pi_{km}} \quad \text{for} \quad j = 1, \ldots, N-1$$

These frequencies will equilibrate at values $p_1^*, \ldots, p_{N-1}^*$, which satisfy the equations

$$p_j^* = (1/\lambda^*) \sum_{k=1}^{N-1} p_k^*\pi_{kj} \quad \text{for} \quad j = 1, \ldots, N-1$$

where

$$\lambda^* = \sum_{k=1}^{N-1}\sum_{m=1}^{N-1} p_k^*\pi_{km}$$

This system of equations can be written more concisely using matrix notation. Let

$$\tilde{\Pi} = (\pi_{ij})_{i,j=1}^{N-1} \quad \text{and} \quad \mathbf{p}^* = (p_1^*, \ldots, p_{N-1})$$

Then $\lambda^*\mathbf{p}^* = \mathbf{p}^*\tilde{\Pi}$. We therefore recognize λ^* as being an eigenvalue of $\tilde{\Pi}$ and $\mathbf{p}^*$ a corresponding eigenvector: λ^* is the largest eigenvalue of the matrix $\tilde{\Pi} = (\pi_{ij})_{i,j=1}^{N-1}$. This follows from the Perron–Froebenius theory for positive matrices. The value of λ^* can be calculated explicitly as

$$\lambda^* = 1 - (1/(2N-1))$$

[see Cannings (1974)], and it can be used directly to calculate the rate of increase of the heteroplasmid population. A starting population of H_0 heteroplasmid cells gives rise to $(2\lambda^*)^n H_0$ after n generations of growth.

Nonselective growth. Here we are interested in the rate at which heteroplasmid cells disappear through segregation during growth under nonselective conditions. As shown in the preceding section, this rate is given by the $\lambda^* = 1 - 1/(2N - 1)$; that is, in the nonselective medium, $(\lambda^*)^n$ represents the proportion of the initial heteroplasmid population (H_0) that is still in the heteroplasmid state (H) after the overall population has undergone n generations of growth in a nonselective medium. In Table 3.2 is listed a series of values for λ^* as a function of copy number N and, for each N, the heteroplasmid fraction after 10 and 30 generations. Note that the value of λ^* can be readily obtained from experimental data since, in a nonselective medium,

$$H = (2\lambda^*)^n H_0 \quad \text{and} \quad T = (2^n) T_0$$

Thus,

$$H/T = (\lambda^*)^n H_0 / T_0$$

where T_0 and T are the total population at the outset and after n generations, respectively.

We denote the equilibrium frequencies that will be approached asymptotically by the population after long-term growth by $p_0^*, \ldots,$ p_n^* under nonselective conditions. These must satisfy the $N + 1$ equations

$$p_j^* = \sum_{i=0}^{N} p_i^* \pi_{ij} \quad \text{for} \quad j = 0, 1, \ldots, N$$

These equations can be solved in terms of the initial population distribution, $\mathbf{p}^{(0)}$. We first observe that

$$\sum_{j=0}^{N} j\pi_{i,j} = i \quad \text{for} \quad i = 0, \ldots, N \qquad \text{(Martingale property)}$$

It follows that

$$p_0^* = p_0^{(0)} + \sum_{i=1}^{N} \left(1 - \frac{i}{N} \right) p_i^{(0)}, \quad p_N^* = 1 - p_0^*, \quad \text{and}$$

$$p_j^* = 0 \quad \text{for} \quad j = 1, \ldots, N - 1$$

Details of this calculation are presented in the discussion of "no selection, no mutation" in Section 3.2B. If the initial distribution is symmetrical, the asymptotic values, p_0^* and p_N^* will, of course, both be 0.5.

Table 3.2 *Effect of copy number on predicted survival of heteroplasmid cells*

Copy number N	Regular–regular model			Random–regular model		
	λ^* $\left(1 - \dfrac{1}{2N-1}\right)$	Heteroplasmid fraction,[a] $H/T[=(\lambda^*)^n]$		μ^* $\left(\dfrac{(N-1)}{(2N-1)}\dfrac{(2N+1)}{(N+1)}\right)$	Heteroplasmid fraction,[a] $H/T[=(\mu^*)^n]$	
		$n = 10$	$n = 30$		$n = 10$	$n = 30$
3	0.8	0.11	0.0012	0.70	0.028	2.3×10^{-5}
4	0.857	0.21	0.0098	0.77	0.075	4.2×10^{-4}
5	0.889	0.31	0.029	0.815	0.13	2.0×10^{-3}
6	0.909	0.39	0.057	0.844	0.18	6.0×10^{-3}
8	0.933	0.50	0.13	0.881	0.28	0.025
10	0.947	0.58	0.20	0.904	0.37	0.045
15	0.966	0.70	0.35	0.935	0.51	0.13
20	0.974	0.77	0.46	0.951	0.61	0.22
30	0.983	0.84	0.60	0.967	0.72	0.37

[a]Fraction of a starting population of heteroplasmid cells that still carries both plasmids after n generations of growth in nonselective medium.

Random replication, regular partitioning. This is a special case of the probability system known as Polya's urn [see Feller (1968, p. 118)]. We derive it here in the context of plasmid replication.

The N plasmids will be replicated during the cell cycle. These replications occur sequentially, as the following example illustrates.

Example. Consider a newborn cell with 1 P'- and 3 P''-plasmids. The copy number is $N = 4$. This cell is denoted by 1|3. Four replications will occur.

First replication: The P'-plasmid is selected for replication once in four times (probability $\frac{1}{4}$); a P''-plasmid is selected three of four times (probability $\frac{3}{4}$). The possibilities are:

(a) 1|3 → 2|3, with probability $\frac{1}{4}$.

(b) 1|3 → 1|4, with probability $\frac{3}{4}$.

Second replication: Following case (a) for the first replication, we have

(a′) 2|3 → 3|3, with probability $\frac{2}{5}$

2|3 → 2|4, with probability $\frac{3}{5}$.

Following case (b), we have

(b′) 1|4 → 2|4, with probability $\frac{1}{5}$

1|4 → 1|5, with probability $\frac{4}{5}$,

and so on.

Next, let us consider full replication sequences.

1|3 → 1|4 → 1|5 → 1|5 → 1|7   with probability $(\frac{3}{4})(\frac{4}{5})(\frac{5}{6})(\frac{6}{7})$

1|3 → 1|4 → 1|5 → 1|6 → 2|6   with probability $(\frac{3}{4})(\frac{4}{5})(\frac{5}{6})(\frac{1}{7})$

and so on.

We must evaluate the probabilities of all possible replications sequences. The sequence 1|3 → ... → 2|6 can occur in $\binom{4}{1}$ ways because P' must be selected at one in four of the replications to have this result. Therefore,

$$\Pr[1|3 \to \ldots \to 2|6] = \binom{4}{1}(\tfrac{3}{4})(\tfrac{4}{5})(\tfrac{5}{6})(\tfrac{1}{7})$$

The probabilities of the other sequences can be constructed similarly. These are summarized as follows:

$$\Pr[1\,|\,3 \rightarrow \ldots \rightarrow 1\,|\,7] = \binom{4}{0}(\tfrac{3}{4})(\tfrac{4}{5})(\tfrac{5}{6})(\tfrac{6}{7})$$

$$\Pr[1\,|\,3 \rightarrow \ldots \rightarrow 2\,|\,6] = \binom{4}{1}(\tfrac{3}{4})(\tfrac{4}{5})(\tfrac{5}{6})(\tfrac{1}{7})$$

$$\Pr[1\,|\,3 \rightarrow \ldots \rightarrow 3\,|\,5] = \binom{4}{2}(\tfrac{3}{4})(\tfrac{4}{5})(\tfrac{1}{6})(\tfrac{2}{7})$$

$$\Pr[1\,|\,3 \rightarrow \ldots \rightarrow 4\,|\,4] = \binom{4}{3}(\tfrac{3}{4})(\tfrac{1}{5})(\tfrac{2}{6})(\tfrac{3}{7})$$

$$\Pr[1\,|\,3 \rightarrow \ldots \rightarrow 5\,|\,3] = \binom{4}{4}(\tfrac{1}{4})(\tfrac{2}{5})(\tfrac{3}{6})(\tfrac{4}{5})$$

Random replication model. The example illustrates the process. The general case can be analyzed in the following way. Given copy number N and a newborn cell having i P'-plasmids, we have

$$
\begin{aligned}
Q_{ik} = {}& \Pr[\text{number of } P' \text{ plasmids after replication} = k\,| \\
& \text{number of } P' \text{ plasmids in newborn} = i] \\
= {}& \binom{N}{k-i} \frac{\eta_{ik}\eta'_{ik}}{N(N+1)\ldots(2N-1)}
\end{aligned}
$$

where

$$
\begin{aligned}
\eta_{ik} &= 1 && \text{if} \quad k = i \\
&= i(i+1)\ldots(k-1) && \text{if} \quad i < k \le N + i \\
&= 0 && \text{otherwise}
\end{aligned}
$$

and

$$
\begin{aligned}
\eta'_{ik} &= 1 && \text{if} \quad 2N - k = N - i \\
&= (N-1)\ldots(2N-k-1) && \text{if} \quad 2N - k > N - i \\
&= 0 && \text{otherwise}
\end{aligned}
$$

The population distribution after random replication and regular partitioning can be determined in a way similar to that in the preceding section. Again, let p_i denote the proportion of the population (after cell division) having i p'- and $N - i$ P''-plasmids. And let p_j^+ denote the proportion of the daughter cells having j P'- and $N - jP''$-plasmids. Then, as before,

$$p_j^+ = \sum_{i=0}^{N} \sum_{k=0}^{2N} p_i Q_{ik} \binom{k}{j} \binom{2N-k}{N-j} \binom{2N}{N}^{-1}$$

for $j = 0, 1, \ldots, N$

Or we can write this more concisely as

$$p_j^+ = \sum_{i=0}^{N} p_i \Pi_{ij}$$

where

$$\Pi_{ij} = \sum_{k=0}^{2N} Q_{ik} \binom{k}{j} \binom{2N-k}{N-j} \binom{2N}{N}^{-1} \quad \text{(Polya chain)} \tag{3}$$

These transition probabilities are defined in a quite complicated way.

Equilibrium in a selecting medium. The matrix $(\Pi_{ij})_{i,j=0}^{N}$ has two eigenvalues equal to 1, corresponding to the two absorbing (null) states, $i = 0$ and $i = N$. The next largest eigenvalue, call it μ^*, gives once again the rate of increase of the heteroplasmid population in a selective medium, $H = H_0(2\mu^*)^n$. This eigenvalue can be calculated directly for this matrix by observing that the row sum of the columns is a constant. A Markov chain having this property is referred to as being *doubly stochastic*. This constant must be the eigenvalue. A short calculation shows that

$$\mu^* = \sum_{i=1}^{N-1} \Pi_{ij} = \frac{N-1}{2N-1} \left(\frac{2N+1}{N+1} \right) \quad \text{for} \quad j = 1, \ldots, N-1$$

The corresponding eigenvector is

$$p^* = \frac{1}{N-1} (0, 1, 1, 1, \ldots, 1, 0)$$

which gives the equilibrium distribution of plasmid types in the population raised in a selecting medium. In particular, after a large number of generations the actual number of cells in each of the classes is approximately:

number of cells having i P'-plasmids

$$= \frac{(2\mu^*)^n}{N-1} \quad \text{for} \quad i = 1, \ldots, N-1$$

Equilibrium in a nonselecting medium. As in the regular–regular case, the eigenvalue (here μ^*) represents the fraction of heteroplasmid organisms that survive cell division. Since $\mu^* < \lambda^*$ for $N > 1$, heteroplasmid cells will segregate more rapidly if replication is random than if it is regular. This is intuitively obvious because random replication will by itself produce substantial disparities in the number of P'- and P''-plasmids in individual cells. These effects are most pronounced for smaller copy numbers, but it is likely that the distinction could be made experimentally even for rather large copy numbers.

After many generations in a nonselecting medium, the frequencies will approach an asymptotic state, where the relative proportion of heteroplasmid cells is negligible. The population's distribution is given (approximately) by

$$\text{number of cells having } N \ P'\text{-plasmids} = 2^n p_0^*$$
$$\text{number of cells having } N \ P''\text{-plasmids} = 2^n p_N^*$$

where the proportions

$$p_0^* = p_0^{(0)} + \sum_{i=1}^{N} \left(1 - \frac{i}{N} \right) p_i^{(0)}$$

and

$$p_N^* = 1 - p_0^*$$

The absorption probabilities therefore are given explicitly in terms of the initial population's distribution, which, in turn, is given by the proportions

$$p_i^{(0)} = \text{proportion of initial population } (n = 0)$$
$$\text{having } i \ P'\text{-plasmids} \quad \text{for} \quad i = 0, 1, \dots, N$$

Again, for symmetrical distributions, $p_0^* = p_N^* = \frac{1}{2}$.

The Markov chain models developed here are valid for all plasmid numbers, and provide explicit solutions for the kinetics of plasmid segregation. The model compares favorably with experimental data for R1/R100 plasmids [see Novick and Hoppensteadt (1978)].

It is perhaps worth noting that the segregation of incompatible plasmids on the basis of random assortment is formally analogous to the random fixation of one of a pair of alleles in a randomly mating finite population of individuals as well as to the segregation of genetically marked organelles in a heteroplasmon.

3.2 Human genetics: Mendelian traits in diploid organisms

Some human cells have single chromosomes (sperm or eggs), but most have chromosomes occurring in matched sets. When a single chromosome occurs, the cell is called a *haploid* cell. It is a *diploid* cell if the chromosomes occur in matched homologous pairs. Human beings are diploid organisms having 23 chromosome pairs, and many plants are polyploids.

Attention is directed first at one gene locus having two alleles. These are denoted by A and a. If the organism is haploid, then it can be either of type A or type a at that locus. Therefore, a population of these organisms is partitioned by this locus into those of type A and those of type a. If the organism is diploid, then the possible types at the locus are AA, Aa, and aa. Note that Aa and aA are indistinguishable in the organism, and so are lumped together in the single notation Aa. These genetic types in a population are called the *genotypes*.

Cell reproduction occurs either through asexual reproduction (mitosis) or sexual reproduction (meiosis). In sexual reproduction, diploid parents each form haploid cells called *gametes*. These are the sperm (male) and the ova (female). The gametes combine to form a fertile cell called the *zygote*. The zygote is a diploid cell that goes on to reproduce by mitosis. The gametes can be thought of as having chromosomes being one strand each from each parent, although the actual situation is more complicated. For example, if the parents have genotypes AA and aa, respectively, then the gametes are A and a, respectively, so the offspring must have genotype Aa.

Cells having genotypes AA and aa are called *homozygotes* and the Aa's are *heterozygotes*. So mating of homozygotes results in homozygous or heterozygous progeny, depending on whether or not the homozygotes are identical. The type of matings and resulting frequencies of progeny genotypes were observed in 1850 by G. Mendel (1865) for randomly mating populations. These observations are summarized in Table 3.3. The laws summarized in this table play a basic role in the models constructed here.

A. *Random sampling for one-locus two-allele traits*

Creation of one generation by its parents can be viewed as a random process. A population of N individuals carries $2N$ genes, and the distribution of their gametes at reproduction time gives the probabilities of various ones being used in reproduction. For example, the offspring of a population consisting of one male and one female [say both heterozygous (Aa) for a one-locus two allele trait] that produce two offspring are summarized in Table 3.4.

This shows that with probability $\frac{1}{8}$ ($= \frac{1}{16} + \frac{1}{16}$) one of the alleles is lost from the population; moreover, with probability $\frac{1}{4}$ ($= \frac{1}{16} + \frac{1}{16} + \frac{1}{8}$) there are no heterozygotes in the population; and so on.

The gene pool of the progeny can be viewed as being formed by

Table 3.3. *Mendel's laws*

| Mating | Progeny | | |
	AA	Aa	aa
$AA \times AA$	1	—	—
$AA \times Aa$	$\frac{1}{2}$	$\frac{1}{2}$	—
$AA \times aa$	—	1	—
$Aa \times Aa$	$\frac{1}{4}$	$\frac{1}{2}$	$\frac{1}{4}$
$Aa \times aa$	—	$\frac{1}{2}$	$\frac{1}{2}$
$aa \times aa$	—	—	1

Table 3.4. *Two offspring of two* Aa's

| Type of offspring | | Probability |
One offspring	The other offspring	
AA	AA	$\frac{1}{16}$
Aa	AA	$\frac{1}{8} + \frac{1}{8} = \frac{1}{4}$
Aa	Aa	$\frac{1}{4}$
aa	Aa	$\frac{1}{4}$
aa	aa	$\frac{1}{16}$
AA	aa	$\frac{1}{8}$

sampling the adult gamete pool. Since the population is finite, this random sampling can have significant effects on the population's genetic structure, particularly if the population is small or the time scale is large, as in evolutionary studies. Many models and methods have been devised for studying these effects. The topics in this section indicate how discrete models are formulated and studied, how they can be analyzed by approximation schemes, and how these results are applied. The development of these topics presented here follows those of Ewens (1969), Crow and Kimura (1970), Moran (1962), Cavalli-Sforza and Bodmer (1971), Ludwig (1974), and Feller (1968).

B. *Fisher–Wright models*

A one-locus two-allele trait carried by a population of N diploid individuals defines a gene pool of size $2N$. Let the alleles be denoted by A and a, and let α_n denote the number of A genes in the nth generation. Thus, $p_n = \alpha_n/2N$ gives the frequency of A genes in the gene pool. The population is assumed to mate at random and be synchronized with nonoverlapping generations. Furthermore, the population size is assumed to remain constant $(= N)$ through the generations.

The sequence of random variables $\{\alpha_n\}$ describes the offspring gene pool immediately after reproduction in successive generations. Wright (1931) and Fisher (1930) studied the dynamic problem by means of the following model. If α_n is known, then α_{n+1} has a binomial distribution with parameter $p_n = \alpha_n/2N$ and index $2N$. Thus, the model is based on sampling of genes with replacement.

No selection, no mutation. Under these conditions, the adult gamete pool at the end of the nth reproduction period has α_n A genes and $2N - \alpha_n$ a genes. Thus,

$$
\begin{aligned}
p_{i,j} = \Pr[\alpha_{n+1} = j \mid \alpha_n = i] &= \binom{2N}{j} p_n^j (1 - p_n)^{2N-j} \\
&= \binom{2N}{j}\left(\frac{i}{2N}\right)^j\left(1 - \frac{i}{2N}\right)^{2N-j} \quad \text{(Fisher–}\\
&\qquad\qquad\qquad\qquad\qquad\qquad\qquad\quad \text{Wright model)}\\
&\text{for}\quad j = 0, 1, \ldots, 2N
\end{aligned}
\tag{4}
$$

Since α_{n+1} is a binomially distributed random variable, it follows that

$$E[\alpha_{n+1}|\alpha_n] = \alpha_n, \qquad \text{var}[\alpha_{n+1}|\alpha_n] = 2Np_n(1 - p_n)$$

[Models other than this have been devised and extensively studied (see Ewens, 1969).]

Let $\phi_{m,n} = \Pr[\alpha_n = m]$. Then the components of the vector $\phi_n = (\phi_{0,n}, \phi_{1,n}, \ldots, \phi_{2N,n})$ give the probability distribution of the gene pool after the nth reproduction period. From the definition of ϕ_n, we see that

$$\phi_{m,n+1} = \Pr[\alpha_{n+1} = m] = \sum_{k=0}^{2N} \Pr[\alpha_n = k]\Pr[\alpha_{n+1} = m|\alpha_n = k]$$

$$= \sum_{k=0}^{2N} \phi_{k,n} p_{k,m}$$

This notation can be simplified by introducing the transition probability matrix $P = (p_{i,j})$ to summarize the transition probabilities. With this, we have $\phi_{n+1} = \phi_n P$, and the solution of this equation is $\phi_n = \phi_0 P^n$, where ϕ_0 is the initial state of the gene pool. For example,

$$\phi_0 = (\delta_{0k}, \delta_{1k}, \ldots, \delta_{2N,k}) \quad \text{where} \quad \delta_{ik} = \begin{cases} 0 & \text{if } i \neq k \\ 1 & \text{if } i = k \end{cases}$$

specifies that the gene pool consists initially of k A genes.

The system's dynamics can therefore be determined by analyzing iterates of the transition probability matrix P. P is a stochastic matrix (i.e., its components are nonnegative and the sum of each row is one), and it has two eigenvalues equal to 1. There are $2N + 1$ possible states for the population each generation, say $S_0, \ldots, S_{2N}$, where S_i corresponds to the gene pool having i A-genes. The state in time period n is determined by α_n. Thus, through the generations, the population goes from state to state with transition probabilities $\{p_{i,j}\}$. The sequence $\{S_{\alpha_n}\}$ describes the states through which the population passes. The sequence $\{\alpha_n\}$ forms a Markov chain; moreover, since $E[\alpha_{n+1}] = \alpha_n$, this is a special Markov chain called a *martingale*.

The interior states $S_1, \ldots, S_{2N-1}$ are transient; for if $p_{i,j}^{(n)}$ indicates the probability of passing from state S_i to state S_j in n steps, then $p_{i,j}^{(n)} \to 0$ as $n \to \infty$ for $k = 1, \ldots, 2N - 1$. To see this, we note that

$$p_{i,j}^{(n)} = (P^n)_{i,j} = (i,j)\text{th element of } P^n$$

Thus, the problem is reduced to a study of the iterates of P.

The eigenvalues of P are

$$\lambda_0 = 1, \quad \lambda_1 = 1, \quad \lambda_2 = 1 - \frac{1}{2N},$$
$$\lambda_j = \frac{2N(2N-1)\ldots(2N-j+1)}{(2N)^j}, \qquad j = 3, \ldots, 2N$$

Using the spectral decomposition of P, we have $P = P_{0,1} + P_2$, where $P_{0,1}$ accounts for projection onto the two unit eigenvalue modes and so has nonzero elements only in the first and last columns, and $P_2^n = O((1 - 1/2N)^n)$. Therefore, $P^n = P_{0,1}^n + P_2^n$ and so approaches a matrix having zero components except in the first and last columns. This shows that $p_{jk}^{(n)} \to 0$ as $n \to \infty$ for $k \neq 0, 2N$.

It follows that for large n

$$\sum_{k=0}^{2N} p_{jk}^{(n)} k \sim 2N p_{j,2N}^{(n)}$$

Since

$$\sum_{k=0}^{2N} p_{jk}^{(n)} k = E[\alpha_{l+n}|\alpha_l = j] = j \quad \text{for any } l \qquad \begin{array}{l}\text{(martingale}\\\text{property)}\end{array}$$

we have that

$$\lim_{n\to\infty} p_{j,2N}^{(n)} = j/2N, \qquad \lim_{n\to\infty} p_{j0}^{(n)} = 1 - j/2N.$$

That is, a population beginning in state S_j has probability $j/2N$ of being absorbed in state S_{2N} and probability $1 - j/2N$ of being absorbed in the state S_0. In addition,

$$\lim_{n\to\infty} P^n = \frac{1}{2N} \begin{bmatrix} 2N & & & 0 \\ 2N-1 & & & 1 \\ 1 & \bigcirc & & 2N-1 \\ 0 & & & 2N \end{bmatrix}$$

This calculation shows that sampling effects will eventually lead to one of the alleles being eliminated from the population! However, if N is large, the approach to fixation of one or the other of the genes is very slow.

These calculations illustrate the phenomenon of *random genetic drift*. We see that if $\alpha_0 = 1$, then the probability is $1/2N$ that the A gene will eventually dominate the gene pool simply as a result of gene sampling at reproduction.

Selection and mutation. Selective and mutational effects complicate this model. The dynamics will be described just as in the preceding section, with the exception that the gamete pool will be sampled at the end of the nth reproductive cycle to form the gene pool in the $(n + 1)$st period. As before, we let α_n denote the number of A genes in the gene pool in the nth generation. Therefore, the expected frequencies of AA, Aa, and aa genotypes in the next generation are

$$(\alpha_n/2N)^2, \quad 2\alpha_n(2N - \alpha_n)/(2N)^2, \quad [(2N - \alpha_n)/2N]^2$$

respectively. If these genotypes have relative viabilities r, s, and 1, respectively, then the expected frequency of A genes at the next reproduction time is

$$(\tilde{p}|\alpha_n) = \frac{r\alpha_n^2 + s\alpha_n(2N - \alpha_n)}{r\alpha_n^2 + 2s\alpha_n(2N - \alpha_n) + (2N - \alpha_n)^2}$$

If in addition to natural selection, A genes mutate to a genes with probability μ and a genes mutate to A genes with probability ν, then the expected frequency of A genes is

$$(p|\alpha_n) = (1 - \mu)(\tilde{p}|\alpha_n) + \nu(1 - (\tilde{p}|\alpha_n))$$

This describes the distribution of the gamete pool at the $(n + 1)$st reproduction time. It follows that

$$p_{ij} = \Pr[\alpha_{n+1} = j \,|\, \alpha_n = i] = \binom{2N}{j}(p|i)^j(1 - (p|i))^{2N-j} \quad (5)$$

Once again $\{\alpha_n\}$ describes a Markov chain; now, however, the martingale property fails, and analysis of the gene pool's dynamics becomes more complicated. Once again, a more sophisticated approach provides a description of the process. This is discussed in Section 5.2B.

C. Branching process approximation: the Galton–Watson–Fisher model

Another point of view can be taken toward the Fisher–Wright model, with interesting results. This approach is in the spirit of *branching processes,* and it focuses on the fate of a small number of A genes.

Suppose that each gene leaves a random number of offspring genes with probability of 0, 1, 2, . . . , being given by the numbers $f_0, f_1, f_2,$ . . . , respectively. The sequence $\{f_j\}$ then gives the probability distribution of offspring. The generating function for this distribution is defined by the formula

$$f(x) = \sum_{j=0}^{\infty} f_j x^j \qquad \text{(generating function for offspring distribution)}$$

We will use this to calculate extinction probabilities. Note that it is assumed that the offspring distribution does not depend on the parents' history.

Again let α_n denote the number of A genes in the nth generation.

Generating function. First,

$$\Pr[\alpha_{n+1} = k | \alpha_n = i] = \sum_{j_1 + \ldots + j_i = k} f_{j_1} \ldots f_{j_i}$$

The conditional generating function is defined by

$$\begin{aligned} F_{n+1,i}(x) &= \sum_{k=0}^{\infty} \Pr[\alpha_{n+1} = k | \alpha_n = i] x^k \\ &= \sum_{k=0}^{\infty} \sum_{j_1 + \ldots + j_i = k} f_{j_1} \ldots {}_{j_i} x^{j_1} \ldots x^{j_i} \\ &= [f(x)]^i \end{aligned}$$

Next,

$$\Pr[\alpha_{n+1} = k] = \sum_{i=0}^{\infty} \Pr[\alpha_{n+1} = k | \alpha_n = i] \Pr[\alpha_n = i]$$

The generating function of $(n + 1)$st generation is defined to be

$$\begin{aligned} F_{n+1}(x) &= \sum_{k=0}^{\infty} \Pr[\alpha_{n+1} = k] x^k \\ &= \sum_{k=0}^{\infty} \sum_{i=0}^{\infty} \Pr[\alpha_{n+1} = k | \alpha_n = i] \Pr[\alpha_n = i] x^k \\ &= \sum_{i=0}^{\infty} \Pr[\alpha_n = i] [f(x)]^i \qquad \text{(generating function for the} \\ &= F_n[f(x)] \qquad \text{Galton–Watson–Fisher process)} \quad (6) \end{aligned}$$

This recursion equation for the generating function of the Galton–Watson process is very useful.

Expected number of A genes. Differentiating this recursion with respect to x gives

$$F_n'(x) = F_{n-1}'[f(x)]f'(x) = \cdots$$
$$= F_0'[f(x)][f'(x)]^n$$

In particular,

$$F_n'(1) = F_0'(1)[f'(1)]^n$$

However,

$$f'(1) = \sum_{j=0}^{\infty} jf_j = \text{expected number of offspring of } A \text{ genes}$$

and $F_n'(1) = $ expected number of A genes in the nth generation. If $f'(1) < 1$, the expected number of A genes approaches zero; if $f'(1) = 1$, this remains constant; and if $f'(1) > 1$, the expected number of A genes grows exponentially.

Probability of extinction of a single mutant. Suppose that a particular locus has only one gene, a. When a single mutation $a \rightarrow A$ occurs, the population consists of all aa's except for one mutant having the Aa genotype. If the fitness of the mutant is s (that is, the expected number of progeny from Aa genotypes in the next generation is s), then

$$f'(1) = \sum_{i=0}^{\infty} if_i = s$$

Thus, $f'(1)$ gives the mutant's fitness.

If a mutation $(a \rightarrow A)$ occurs at time $n = 0$, then $\alpha_0 = 1$. Let ϵ be the probability of eventual extinction of the A gene. If the mutant leaves j offspring, then the probability of eventual extinction of all j progeny is ϵ^j. Therefore,

$$\epsilon = \Pr[\text{eventual extinction}]$$
$$= \sum_{j=0}^{\infty} \Pr[j \text{ offspring}]\Pr[\text{all } j \text{ become extinct}]$$
$$= \sum_{j=0}^{\infty} f_j \epsilon^j = f(\epsilon)$$

and the equation

(extinction probability) $\epsilon = f(\epsilon)$

determines the extinction probability.

Iteration of the function $f(x)$ is equivalent to solving the difference equation $x_{n+1} = f(x_n)$, because the solution of both is $x_n = f^{(n)}(x_0)$. Now, the function $f(x)$ and its derivatives are nonnegative, so in particular, the function is convex. If $f'(x) \leq 1$, there is only one static state in the unit interval: $x = 1$. In this case, $\epsilon = 1$, so extinction is certain. If $f'(1) > 1$, there are two static states: $x = \epsilon < 1$ and $x = 1$.

We have now seen that if $f'(1) \leq 1$, then $F_n(x) = f^{(n)}(x) > 1$, whereas if $f'(1) > 1$, $F_n(x) \to \epsilon$ as $n \to \infty$. Note that in any case

$F_1(0) = \Pr[\alpha_1 = 0] =$ probability of extinction of first generation

$$F_2(0) = \sum_{j=0}^{\infty} f_j(f(0))^j = \left. \sum_{j=0}^{\infty} \sum_{i=0}^{\infty} \Pr[\alpha_2 = j | \alpha_1 = i]\Pr[\alpha_1 = i]x^j \right|_{x=0}$$

= probability of extinction by second generation

$F_n(0) =$ probability of extinction by nth generation

If $\{x_n\}$ is defined by $x_{n+1} = f(x_n)$, $x_0 = 0$, we see that $x_n = F_n(0)$. As before, $x_n \to \epsilon$.

Probability of fixation. Next, we consider the probability of fixation, $\phi = 1 - \epsilon$. Here ϕ satisfies

$$\phi = 1 - f(1 - \phi) = 1 - f(1) + f'(1)\phi$$
$$- f''(1)\phi^2/2 + 0(\phi^3) \sim s\phi + (\sigma^2 + s^2)\phi^2/2$$

where σ^2 is the variance of the distribution of offspring genes f_j. If $\phi \ll 1$, we see that

$$\phi \sim 2(s - 1)/(\sigma^2 + s^2)$$

if $s > 1$. (Recall that $s \leq 1$ implies that $\phi = 0$.)

Poisson distribution of offspring. Finally, suppose that the offspring have a Poisson distribution. If a mutant has fitness s, then $f(x) = e^{s(x-1)}$. The probability of eventual extinction, ϵ, is determined from the equation $\epsilon = e^{s(\epsilon-1)}$. Moreover, the probability ϕ that the A gene is eventually fixed in the population is $\phi = 1 - \epsilon$. This is determined by $\phi = 1 - e^{-s\phi}$, so $\phi \sim 2(s - 1)s^2$ if $s \sim 1$. Many other calculations can be carried out with the Galton–Watson model (see Harris, 1963).

3.3   Contagion

Epidemics are natural phenomena of immense proportions. Plague in the fourteenth century is calculated to have killed an estimated 25 million people, and in 1918 influenza reached 50 percent of the world's population and killed about 20 million people. Today, malaria and schistosomiasis are the most widely spread contagious diseases in human beings, although influenza, plague, cholera, measles, and many others are significant. The economic impact alone of these diseases is staggering, with workdays lost, certain areas being uninhabitable, and so on.

Epizootics, diseases in animals, although less well known, certainly have had similar severe outbreaks. These have often had important consequences for human beings because animals frequently act as intermediate and primary hosts for diseases to which human beings are susceptible. In fact, schistosomiasis can be viewed as a disease of snails and plague one of rats. Also important are diseases in food-producing animals such as cattle, poultry, and fish.

The study of diseases and their propagation in populations is complicated for several reasons. The etiology of disease in an individual organism is difficult to trace, and is frequently only guessed at. The detection of disease in an organism is often difficult because of false test results, carrier effects, and definition alone: When does a susceptible organism become infectious? Finally, good-quality data are very difficult to collect because of inaccessibility of the populations, even human beings.

The Reed–Frost model is basic in the theory of epidemics.

A. *Reed–Frost model: S → I → R*

A model of susceptible–infective interaction was introduced by L. J. Reed and W. H. Frost in 1928 (see Bailey, 1957). First, an appropriate time unit is selected; usually, this is related to the mean length of the infectious period of the disease. Then the number of susceptibles and infectives in each period can be described. Let $\mathcal{S}_n$ = number susceptible in nth period, $\mathcal{I}_n$ = number infective in nth period, and $\mathcal{R}_n$ = number of those removed in nth period. That is,

$\mathscr{R}_n$ denotes the number of individuals who have passed through the infectious stage by the nth period.

The population is assumed to be constant and randomly mixing. That is, the probability of effective contact between any susceptible and any infective is the same for the entire population. Let p = probability of an effective contact between a susceptible and an infective in the nth time period. At the end of each time period, the population will be examined and the new numbers of susceptibles, infectives, and removed will be determined. The probability that a given susceptible will avoid contact with all infectives is $(1 - p)^{\mathscr{J}_n}$ = probability of a given susceptible not contacting any infectives in the nth time period. Therefore,

$$\Pr[\mathscr{S}_{n+1} = k, \mathscr{J}_{n+1} = m \,|\, \mathscr{S}_n, \mathscr{J}_n]$$

$$= \binom{\mathscr{S}_n}{k}(1 - p)^{\mathscr{J}_n k}[1 - (1 - p)^{\mathscr{J}_n}]^m \qquad \begin{array}{l}\text{(Reed--} \\ \text{Frost model) (7)}\end{array}$$

That is, the probability that $\mathscr{S}_{n+1} = k$ and $\mathscr{J}_{n+1} = m$, given $\mathscr{S}_n$ and $\mathscr{J}_n$, is the number of ways k individuals can be selected from among $\mathscr{S}_n$, times the probability that k miss contact and m make contact. If $\mathscr{S}_{n+1} = k$ and $\mathscr{J}_{n+1} = m$, then $\mathscr{R}_{n+1} = N - k - m$, because the total population remains constant.

Since

$$E(\mathscr{S}_{n+1} \,|\, \mathscr{S}_n, \mathscr{J}_n) = \mathscr{S}_n(1 - p)^{\mathscr{J}_n} \quad \text{and}$$
$$E(\mathscr{J}_{n+1} \,|\, \mathscr{S}_n, \mathscr{J}_n) = \mathscr{S}_n(1 - (1 - p)^{\mathscr{J}_n})$$

we are motivated to consider the system of equations

$$S_{n+1} = S_n(1 - p)^{I_n}, \qquad I_{n+1} = S_n[1 - (1 - p)^{I_n}]$$

as a deterministic analog of the Reed--Frost model. However, we cannot expect that $E(\mathscr{S}_n) = S_n$ and $E(\mathscr{J}_n) = I_n$. It is convenient to rewrite this deterministic system as

$$S_{n+1} = \exp(-aI_n)S_n, \qquad I_{n+1} = S_n(1 - \exp(-aI_n)) \qquad (8)$$

where $a = -\log(1 - p)$. This is called the Kermack--McKendrick (1927) model and it is studied in Section 4.2A. The Reed--Frost model defines $\mathscr{S}_n$ and $\mathscr{J}_n$ as a binomial Markov chain in which the transition probability changes with time depending on the epidemic's state. This chain is difficult to analyze, so we turn to computer simulations.

B. *Monte Carlo simulation*

The Reed–Frost model just introduced was given in terms of random variables $\mathcal{S}_n$, $\mathcal{J}_n$, and $\mathcal{R}_n$, which specify the actual numbers of susceptibles, infectives, and removals in the nth time period. Recall that the population size was fixed, $\mathcal{S}_n + \mathcal{J}_n + \mathcal{R}_n = N$, and the conditional density functions of $\mathcal{S}_n$ and $\mathcal{J}_n$ are given recursively by the formulas

$$\Pr[\mathcal{S}_{n+1} = k, \mathcal{J}_{n+1} = m \mid \mathcal{S}_n, \mathcal{J}_n]$$
$$= \binom{\mathcal{S}_n}{k}(1 - p)^{\mathcal{J}_n k}(1 - (1 - p)^{\mathcal{J}_n})^m, \qquad k + m = \mathcal{S}_n$$

where p is the probability of an effective contact in time period n.

This model can be analyzed by (a) calculating the joint probability density function

$$\Pr[\mathcal{S}_n = k, \mathcal{J}_n = m]$$

directly, by (b) computer simulation of the model, or (c) by various approximation schemes. Although direct calculation of the joint density function is feasible, it is tedious and the results are of limited value. On the other hand, some striking results have been found by computer simulation.

A Monte Carlo simulation of the Reed–Frost chain proceeds by generating a number of possible evolutions of the epidemic using a computer; these are called *sample paths*. Various statistics of a collection of sample paths can be calculated to describe the epidemic's dynamics.

Calculation of a sample path. We begin with initial data, say $\mathcal{S}_0 = 40$ and $\mathcal{J}_0 = 1$, $\mathcal{R}_0 = 0$. The Reed–Frost model then states that

$$Q_{1,k} = \Pr[\mathcal{S}_1 = k, \mathcal{J}_1 = 41 - k \mid \mathcal{S}_0 = 40, \mathcal{J}_0 = 1]$$
$$= \binom{40}{k} q^k (1 - q)^{41-k}$$

where $q = 1 - p$. The computer is instructed to generate a random number from a uniform distribution on the interval $[0, 1]$, say x_1. Next, the numbers $Q_{40}, Q_{39}, \ldots, Q_j$ are calculated until $Q_{40} + Q_{39} + \ldots + Q_{j_1} > x_1$. Then we take $\mathcal{S}_1 = j_1$, $\mathcal{J}_1 = 41 - j_1$; another random variable, x_2, is generated, and the numbers

$$Q_{2,k} = \Pr[\mathcal{S}_2 = k, \mathcal{I}_2 = j_1 - k \,|\, \mathcal{S}_1 = j_1, \mathcal{I}_1 = 41 - j_1]$$

$$= \binom{j_1}{k} q^{\mathcal{I}_1 k} (1 - q^{\mathcal{I}_1})^{j_1 - k}$$

are calculated $(Q_{2,j_1}, Q_{2,j_1-1}, \ldots, Q_{2,j_2})$ until $Q_{2,j_1} + \ldots + Q_{2,j_2} > x_2$. Then $\mathcal{S}_2 = j_2$, $\mathcal{I}_1 = j_1 - j_2$. This process is continued until a number of generations, say 100, are calculated. The result is the sample path $j_1, j_2, \ldots, j_{100}$ describing the susceptible population sizes.

A faster method for generating sample paths takes advantage of the binomial distribution. At each n, we generate j_{n-1} random numbers $\rho_1, \ldots, \rho_{j_{n-1}}$. We let $\sigma_i = 1$ if $\rho_i \leq q$, and $\sigma_i = 0$, otherwise. Then $j_n = \sum_{i=\phi}^{j_{n-1}} \sigma_i$.

This is repeated, say 50 times, and a collection of 50 numbers $\{j_{100}\}$ are collected. A number of techniques can be used to process this collection. For example, a histogram can be plotted. A typical case is shown in Table 3.5. Such a histogram describes the final size distribution of the susceptible population.

Monte Carlo methods are used to study a wide variety of problems (see Hammersley and Handscomb, 1967).

3.4   Summary

Several Markov chain models of population phenomena have been presented in this chapter. The two-state chain (1), the hypergeometric (2) and Polya (3) chains, The Fisher–Wright chains [(4) and (5)], and the Reed–Frost chain (7) are all studied by iteration of the corresponding transition probability matrix. In special cases of martingales [(1)–(4)], the asymptotic behavior (large n) can be determined simply by correctly projecting the initial data onto the dominant eigenvectors. The problem is more difficult for the other chains presented here.

Table 3.5. *Histogram of the final size distribution of the susceptible population*

j_{100}	0	1	2	3	...	34	35	36	37	38	39	40
Number of sample paths having this as j_{100}	4	2	1	0	...0...	0	1	3	10	19	8	2

Certain aspects of behavior of the other chains can be indicated by using other methods, such as branching processes (6), Monte Carlo simulations and by deterministic approximations such as the Kermack–McKendrick model (8). Another approximation that is useful is the diffusion approximation, which is described briefly in Section 5.2B. However, these approximations do not answer all questions about Markov chain models, and one is often left with no option other than spectral analysis of the transition matrix. These and additional topics on Markov chains are discussed in Karlin (1966, 1972) and Isaacson and Madsen (1976).

Exercises

3.1 Let P be the transition matrix for the no-selection, no-mutation, Fisher–Wright model:

$$P = (P_{ij})_{i,j=0,\ldots,2N}$$

where

$$P_{ij} = \binom{2N}{j}\left(\frac{i}{2N}\right)^{j}\left(1 - \frac{i}{2N}\right)^{2N-j}$$

Show that the eigenvalues of P are given by

$$\lambda_0 = 1, \qquad \lambda_j = \frac{2N(2N-1)\ldots(2N-j+1)}{(2N)^j},$$

$$j = 1, \ldots, 2N$$

3.2 In the Kermack–McKendrick model, we suppose that $a_n = \epsilon r$, $\alpha_n = 1 - \epsilon q$. Let $S_n = S(n\epsilon)$, $I_n = I(n\epsilon)$, and $R_n = R(n\epsilon)$ where S, I, and R are smooth functions.
(a) Show that $dS/dt = -rIS$, $dI/dt = rIS - qI$, and $dR/dt = qI$.
(b) Solve this system for I as function of S.
(c) *(Threshold theorem)*. Show that $S(\infty)$ exists and obtain a formula for it.
(d) Calculate the final size of a Kermack–McKendrick epidemic as a function of $\gamma = q/r$ with $I_0/S_0 = .001, .01, .1$. Interpret the results.
3.3 *(Epidemics as birth–death processes)*. Let the state of an epidemic at time t be described by random variables $(S(t), I(t), R(t))$, the numbers of susceptible, infective, and removed. The total population is held constant: $S + I + R = N$. Define

$$P_{i,r}(t) = \Pr[I(t) = i, R(t) = r]$$

We suppose that changes of state have the following properties:

$$\Pr[I(t + dt) - I(t) = 1 \mid i, r] = bi(N - i - r)\,dt + o(dt)$$
$$\Pr[\,|I(t + dt) - I(t)| > 1 \mid i, r] = o(dt)$$

$$\Pr[R(t + dt) - R(t) = 1 | i, r] = id\, dt + o(dt)$$
$$\Pr[|R(t + dt) - R(t)| > 1 | i, r] = o(dt)$$

(a) Show that $P_{i,r}(t + dt) = P_{i,r}(t)[1 - bi(N - i - r)\, dt - id\, dt]$
$+ P_{i-1,r}(t)b(i - 1)(N - i - r)\, dt + P_{i+1,r-1}(t)d(i+1)\, dt + o(dt)$

(b) Let $B = Nb$, $\gamma = d/B$, and $t' = Nbt$. Let $dt \to 0$ and show that

$$\frac{dP_{i,r}}{dt} = - \left(i \left(\frac{N - i - r}{N} \right) + \gamma i \right) P_{i,r}$$
$$+ (i - 1) \left(\frac{N - i + 1 - r}{N} \right) P_{i-1,r} + \gamma(i + 1) P_{i+1,r-1}$$

(c) Calculate (approximately) the possibility of extinction in the early stages of the epidemics. In this case, $N \gg i$, $N \gg r$, so

$$dP_{i,r}/dt = -(1 + \gamma)iP_{i,r} + (i - 1)P_{i-1,r} + \gamma(i + 1)P_{i+1,r-1}$$

Summing this formula over r, derive an equation for $P_i = \sum_{r=0}^{\infty} P_{i,r}$.

(d) Let $u(x, t) = \sum_{i=0}^{\infty} P_i(t)x^i$ denote the probability generating function of $I(t)$. Derive a first-order partial differential equation for u. Solve this equation by the method of characteristics, and interpret your answer. [*Hint:* $u(x, t) = \gamma(1 - x) + (x - \gamma)\exp(-(1 - \gamma)t)/(1 - x + (x - \gamma)e^{-(1-\gamma)t})$.]

(e) Show that $\lim_{T \to \infty} P_0(T) = \gamma$. [*Hint:* Use $u(x, t)$.]

(f) Relate this answer to the simulation in Exercise 3.2(d).
Comment: $(I(t), R(t))$ is a Poisson process. It has been referred to as the general stochastic epidemic. It can be also approximated by a diffusion process. For this and other approximations, see Ludwig (1974, pp. 42ff.).

4

Perturbation methods

As we have seen in the Fibonacci model (Section 2.1), a closed form for a model's solution does not necessarily provide useful information; additional analysis of the solution may be required. In that example, we found the approximate behavior of the population's birth rate after a large number of generations ($n \geq 5$) and made precise the sense in which the approximation is made. In many problems, for example those arising in genetics, it is not even possible to obtain a formula for the solution. However, when detailed information is available for certain combinations of parameters, it is frequently possible to derive useful information for nearby values. These are referred to as *perturbation methods*. Examples from genetics and epidemics will be used here to present several perturbation methods that are basic tools in applied mathematics. These problems also focus on deducing the evolution and long-time behavior of biological systems.

4.1 Approximations to dynamic processes: multiple-time-scale methods

In most biological and physical systems, various aspects of a phenomenon take place on various time scales. For example, a population's genetic structure changes as a result of sampling of the gene pool at reproduction times, so these changes occur on a fast time scale compared to infrequent mutations and slight differences between fertility and survivability among genotypes, which are significant only over very long (or slow) time scales.

The most widely used multiple-scale methods fall roughly into two classes: those for which rapid changes occur initially followed by slow long-time-scale changes, and those for which there is persistent rapidly oscillating behavior that is modulated by slowly changing components. These are generally referred to as *matching* and *averaging* methods, respectively. Examples from human genetics will be derived and used to illustrate both of these methods.

A. *Fisher–Wright–Haldane model: the method of averaging*

One-locus two-allele genetic traits were studied in Section 3.2. The population carries a gene pool consisting of A and a genes. The proportion of the gene pool of type A immediately preceding the nth reproduction will be denoted by g_n. The probability of survival to the next reproduction and their fertility will be denoted by r_n, s_n, and t_n for the AA, Aa, and aa genotypes, respectively. Table 4.1 summarizes the notation.

The gene pool is modeled from generation to generation by the sequence $\{g_n\}$, which is, in turn, determined from the equation (see Section 3.2B)

$$g_{n+1} = \frac{r_n g_n^2 + s_n g_n(1 - g_n)}{r_n g_n^2 + 2s_n g_n(1 - g_n) + t_n(1 - g_n)^2} \tag{1}$$

This model will be used to illustrate the method of averaging.

r_n, s_n, t_n *constant.* Let $r_n = r$, $s_n = s$, and $t_n = t$, so that all the fitnesses remain constant through the generations. Then the qualitative behavior of this model can be determined by geometric iteration. In this case

$$g_{n+1} = \frac{r g_n^2 + s(1 - g_n)g_n}{r g_n^2 + 2s g_n(1 - g_n) + t(1 - g_n)^2}$$

Figure 4.1 shows the four possible cases for this equation. This shows that if $r > s > t$, then $g_n \to 1$, so A eventually dominates the gene pool, but if $r < s < t$, then $g_n \to 0$, so A eventually disappears. The remaining two cases are interesting. First, if $s > r$ and $s > t$, then $g_n \to g^*$. Thus, natural selection of genotypes acts to maintain both alleles in the population. The sickle-cell trait is a well-documented case of this: AA's have normal red blood cells but are susceptible to malaria; aa's have badly deformed cells that do not carry oxygen well, resulting in anemia; and the heterozygotes Aa do not suffer from anemia and enjoy some immunity to malaria. The remaining case, $s < r$, $s < t$, is known as *disruptive selection:* If $g_0 < g^*$, then $g_n \to 0$, but if $g_0 > g^*$, then $g_n \to 1$. This phenomenon is not so well documented, but it is believed to play a role in speciation.

Multitime method of averaging. The method of geometric iteration does not work if the fitnesses change from generation to generation.

Table 4.1. *Survival of genetic traits in the Fisher–Wright–Haldane model*

	*n*th reproduction		(*n* + 1)st reproduction
Gene distribution in the reproduction gene pool	A $\quad g_n$		$g_{n+1} = \dfrac{r_n g_n^2 + s_n g_n(1 - g_n)}{r_n g_n^2 + 2s_n g_n(1 - g_n) + t_n(1 - g_n)^2}$
	a $\quad 1 - g_n$		$1 - g_{n+1}$
Genotype distribution in the population	AA	$D_n = g_n^2 N_n$	$r_n D_n$ $\qquad\qquad D_{n+1} = g_{n+1}^2 N_{n+1}$
	Aa	$2H_n = 2g_n(1 - g_n)N_n$	$2s_n H_n$ $\qquad\qquad 2H_{n+1} = 2g_{n+1}(1 - g_{n+1})N_{n+1}$
	aa	$R_n = (1 - g_n)^2 N_n$	$t_n R_n$ $\qquad\qquad R_{n+1} = (1 - g_n)^2 N_{n+1}$

Note: g_n = proportion of the reproductive gene pool of type A immediately preceding the *n*th reproduction
D_n = number of newborn AA genotypes at the *n*th reproduction
$2H_n$ = number of newborn Aa genotypes at the *n*th reproduction
R_n = number of newborn aa genotypes at the *n*th reproduction
N_n = newborn population size = $D_n + 2H_n + R_n$
$(r_n g_n^2 + s_n g_n(1 - g_n))N_n$ = number of A genes participating in the $(n + 1)$st reproduction
$(r_n g_n^2 + 2s_n g_n(1 - g_n) + t_n(1 - g_n)^2)N_n$ = number of genes in the reproductive gene pool preceding the $(n + 1)$st reproduction

However, the method of averaging applies if the sequences $\{r_n, s_n, t_n\}$ are oscillatory. For example, if the generation time is much less than one year, seasonal changes can have an effect on the population's genetic structure.

To investigate this, we suppose that the fitnesses are almost constant but periodic, say with period T:

$$r_n = 1 + \epsilon\rho_n, \qquad s_n = 1 + \epsilon\sigma_n, \qquad t_n = 1 + \epsilon\tau_n$$

Figure 4.1. Gamete frequency curves with various fitnesses. (*a*) Fixation of A: $r > s > t$; $g_n \to 1$. (*b*) Heterosis: $s > r > t$; $g_n \to g^*$. (*c*) Disruptive selection: $t, r > s$; $g_0 > g^*$ implies that $g_n \to 1$, $g_0 < g^*$ implies that $g_n \to 0$. (*d*) Fixation of a, $t > s > r$; $g_n \to 0$.

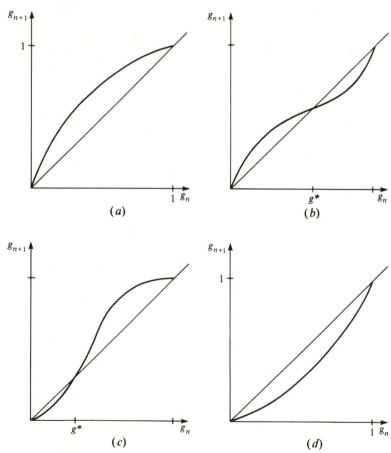

where

$$\rho_{n+T} = \rho_n, \sigma_{n+T} = \sigma_n \quad \text{and} \quad \tau_{n+T} = \tau_n \quad \text{for all } n$$

and where $0 < \epsilon \ll 1$ is a small parameter. The model (1) becomes

$$g_{n+1} = g_n + \epsilon[\sigma_n - \tau_n + g_n(\rho_n - 2\sigma_n + \tau_n)] g_n(1 - g_n) + O(\epsilon^2)$$
$$(2)$$

where $O(\epsilon^2)$ indicates terms that are multiplied by ϵ^2 and higher powers of ϵ.

The multitime method proceeds by replacing g_n with $g_n = G(n, \epsilon n, \epsilon)$ where G is a smooth function of its variables. Substituting this into (2) gives

$$G(n + 1, s + \epsilon, \epsilon) = G(n, s, \epsilon)$$
$$+ \epsilon[\sigma_n - \tau_n + G(n, s, \epsilon)(\rho_n - 2\sigma_n + \tau_n)](1 - G)G$$

where we have set $s = \epsilon n$. The thinking is that two time scales are indicated by (2): n and a slower time scale on which selection acts, $s = \epsilon n$. If these are correctly accounted for, the solution should depend smoothly on ϵ:

$$G(n, s, \epsilon) = G_0(n, s) + \epsilon G_1(n, s) + O(\epsilon^2)$$

The coefficients, $G_0, G_1, \ldots$, can be found by differentiating the equation successively with respect to ϵ. The results are

$$G_0(n + 1, s) = G_0(n, s)$$
$$G_1(n + 1, s) + \frac{dG_0}{ds}(n, s) = G_1(n, s)$$
$$+ [\sigma_n - \tau_n + (\rho_n - 2\sigma_n + \tau_n)] G_0(n, s)(1 - G_0(n, s))$$

The first equation shows that G_0 is independent of n, so we write $G_0 = G_0(s)$.

The second equation can be solved for G_1 by successive back substitutions

$$G_1(n, s) = G_1(0, s)$$
$$+ \sum_{k=0}^{n-1} (\sigma_k - \tau_k + (\rho_k - 2\sigma_k + \tau_k)G_0)(1 - G_0)G_0 - n\frac{dG_0}{ds}$$

If the expansion is valid, then G_1 is bounded for large n. Assuming this to be the case, dividing this equation by n and passing to the limit $n = \infty$, we see that

$$\frac{dG_0}{ds} = (\bar{\sigma} - \bar{\tau} + (\bar{\rho} - 2\bar{\sigma} + \bar{\tau})G_0)G_0(1 - G_0)$$

where

$$\bar{\rho} = \lim_{n \to \infty} (1/n) \sum_{k=0}^{n-1} \rho_k = (1/T) \sum_{k=0}^{T-1} \rho_k,$$

$$\bar{\sigma} = \lim_{n \to \infty} (1/n) \sum_{k=0}^{n-1} \sigma_k = (1/T) \sum_{k=0}^{T-1} \sigma_k,$$

$$\bar{\tau} = \lim_{n \to \infty} (1/n) \sum_{k=0}^{n-1} \tau_k = (1/T) \sum_{k=0}^{T-1} \tau_k$$

It is shown in Hoppensteadt and Miranker (1977) that $g_n = G_0(\epsilon n) + O(\epsilon)$ for $0 \le n \le O(1/\epsilon)$. Table 4.2 shows the behavior of the gene-pool frequencies. Therefore, the mean values of the fitnesses determine the gene pool's evolution.

Table 4.2. *Behavior of gene-pool frequencies in the multitime method*

$\bar{\sigma} - \bar{\tau}$	$\bar{\rho} - 2\bar{\sigma} + \bar{\tau}$	$G_0(\infty)$
+	−	$G_0(\infty) = G^*$
−	−	$G_0(\infty) = 0$
−	+	$\begin{cases} G_0(0) < G^* \Rightarrow G_0(\infty) = 0 \\ G_0(0) > G^* \Rightarrow G_0(\infty) = 1 \end{cases}$
+	+	$G_0(\infty) = 1$

Note: $G^* = |(\bar{\sigma} - \bar{\tau})/(\bar{\rho} - 2\bar{\sigma} + \bar{\tau})|$.

Table 4.3. *Genotype distribution of a one-locus, three-allele genetic trait*

Genotype	AA	AA'	Aa	$A'A'$	$A'a$	aa
Frequency	g_n^2	$2g_n g_n'$	$2g_n g_n''$	$(g_n')^2$	$2g_n' g_n''$	$(g_n'')^2$
Fitness	r_1	r_2	r_3	r_4	r_5	r_6

B. *One-locus three-allele traits under slow selection: the method of matched asymptotic expansions*

Geometric iteration is not useful in dealing with more complicated genetic structures such as the ABO system of blood types. A method called the *method of matched asymptotic expansions* is useful for studying slow selection in such systems.

A model of one-locus three-allele traits is constructed as before. There are now three gamete types making up the gamete pool at each reproduction time, say A, A', and a. Let g_n = frequency of A gametes, and g'_n = frequency of A' gametes. Then the frequency of the a gametes is given by $1 - g_n - g'_n$. The numbers of genotypes resulting in the next generation are summarized in Table 4.3. It follows that

$$g_{n+1} = \frac{r_1 g_n^2 + r_2 g_n g'_n + r_3 g_n (1 - g_n - g'_n)}{w(g_n, g'_n)}$$

$$g'_{n+1} = \frac{r_2 g_n g'_n + r_5 g'_n (1 - g_n - g'_n) + r_4 (g'_n)^2}{w(g_n, g'_n)}$$

where the mean fitness w is given by

$$w(g_n, g'_n) = r_1 g_n^2 + 2r_2 g_n g'_n + 2r_3 g_n g''_n + r_4 (g'_n)^2 + 2r_5 g'_n g''_n + r_6 (g''_n)^2$$

where $g''_n = 1 - g_n - g'_n$.

This system is rather complicated, and instead of presenting a detailed analysis of it, a method is introduced that gives a straightforward description of the gamete frequencies in the special case of slow selection.

If the fitnesses of the various genotypes are nearly identical, then natural selection is acting on a slow time scale. Let

$$r_i = 1 + \epsilon \rho_i \quad \text{for} \quad i = 1, \dots, 6$$

where $\epsilon \ll 1$ and $\epsilon \rho_i \ll 1$. With this change in the model's parameters, a rearrangement leads to

$$g_{n+1} = \frac{g_n + \epsilon(\rho_1 g_n^2 + \rho_2 g_n g'_n + \rho_3 g_n (1 - g_n - g'_n))}{1 + \epsilon \tilde{w}(g_n, g'_n)}$$

$$g'_{n+1} = \frac{g'_n + \epsilon(\rho_2 g_n g'_n + \rho_5 g'_n (1 - g_n - g'_n) + \rho_4 (g'_n)^2)}{1 + \epsilon \tilde{w}(g_n, g'_n)}$$

where

$$\tilde{w}(g_n, g_n') = \rho_1^2 g_n^2 + 2\rho_2 g_n g_n' + 2\rho_3 g_n g_n''$$
$$+ \rho_4(g_n')^2 + 2\rho_5 g_n' g_n'' + \rho_6(g_n'')^2$$

A method for analyzing systems of this kind was developed by Hoppensteadt and Miranker (1977). The result is that these difference equations look like they are forward Euler approximations to differential equations for smooth functions $G(s)$ and $G'(s)$. We write

$$g_n = G(\epsilon n) + O(\epsilon) \quad \text{and} \quad g_n' = G'(\epsilon n) + O(\epsilon)$$

where the functions G and G' are determined by the differential equations

$$dG/dt = (\rho_3 + (\rho_1 - \rho_3)G + (\rho_2 - \rho_3)G')G - \tilde{w}(G, G')G$$
$$dG'/dt = (\rho_5 + (\rho_2 - \rho_5)G + (\rho_4 - \rho_5)G')G' - \tilde{w}(G, G')G'$$

and where

$$\tilde{w} = G^2(\rho_1 + \rho_6 - 2\rho_3) + 2GG'(\rho_2 + \rho_6 - \rho_3 - \rho_5)$$
$$+ (G')^2(\rho_4 + \rho_6 - 2\rho_5) + 2G(\rho_3 - \rho_6) + 2G'(\rho_5 - \rho_6) + \rho_6$$

This system can be analyzed by standard-phase plane techniques.

One-locus three-allele traits arise frequently in blood group studies. The ABO system in human blood (McKusick, 1969) and other systems in baboon blood (Jolly and Brett, 1973) are two well-known examples.

4.2 Static states: bifurcation methods

The appearance of multiple static states for systems is referred to as a *bifurcation* phenomenon. We study here two examples.

A. *Threshold of susceptible population size*

The threshold phenomenon of epidemics is illustrated with the Kermack–McKendrick model (Section 3.3A). We consider model (8) of Chapter 3 in a more general form

$$S_{n+1} = \exp(-aI_n)S_n$$
$$I_{n+1} = \alpha I_n + (1 - \exp(-aI_n))S_n \tag{3}$$
$$R_{n+1} = (1 - \alpha)I_n + R_n$$

Here α is the probability that an infective survives one time period as an infective. If the time periods are chosen equal to the length of the infective period, then $\alpha = 0$ as in (8) of Chapter 3. Of course, only S_n and I_n need be considered because they determine R_n.

First, we consider the initial change in the infective population:

$$\delta I_1 = I_1 - I_0 = (\alpha - 1)I_0 + (1 - \exp(-aI_0))S_0$$

If this quantity is negative, the infection is dying out of the population, and since there is no mechanism for reintroduction of it, I_n is expected to approach zero. On the other hand, if this quantity is positive, we see that each infective is more than replacing himself, and a snowballing effect is expected with an epidemic resulting. The following theorem makes this intuitive argument precise.

Threshold theorem. Suppose that $R_0 = 0$. Then:
  (a) The susceptible population approaches a limiting value;
       $S_n \to S_\infty$ as $n \to \infty$,
  (b) If $F = S_\infty/S_0$, then

$$F = \exp\left(- \frac{aS_0}{1 - \alpha} \left(1 + \frac{I_0}{S_0} - F \right) \right)$$

For I_0/S_0 small, $T = (1 - \alpha)/a$ is the threshold of susceptible population size. If $S_0 < T$, then $F \sim 1$ and few susceptibles are infected. If $S_0 > T$, an epidemic results. [This is explained after the proof.]

Proof of the threshold theorem. (a) As

$$S_{n+1} = \exp(-aI_n)S_n \le S_n$$

and $S_n \ge 0$, for all n, we see that $\lim(n\to\infty)S_n$ exists. Denote this by S_∞.

  (b) As $R_{n+1} = (1 - \alpha)I_n + R_n \ge R_n$ and $R_n \le N$, we see that $\lim(n\to\infty)R_n$ exists; call it R_∞. As $I_n = R_{n+1} - R_n$, we have that $\lim(n\to\infty)I_n = 0$ and so $R_\infty = N - S_\infty$.

  (c) Finally, as

$$S_{n+1} = \exp(-aI_n)S_n$$

we have that

$$S_n = S_0 \exp\left(-a \sum_{k=0}^{n-1} I_k \right)$$

$$= S_0 \exp\left(-\left(\frac{a}{1-\alpha}\right) \sum_{k=0}^{n-1} (R_{k+1} - R_k) \right)$$

$$= S_0 \exp\left(-\left(\frac{a}{1-\alpha}\right) R_{n-1} \right)$$

Passing to the limit $n \to \infty$, we see that

$$S_\infty = S_0 \exp\left(-\left(\frac{a}{1-\alpha}\right)(N - S_\infty) \right)$$

Setting $F = S_\infty/S_0$ gives

$$F = \exp\left(-\left(\frac{aS_0}{1-\alpha}\right)\left(1 + \frac{I_0}{S_0} - F\right) \right)$$

This completes the proof of the theorem.

F is a measure of the epidemic's final size and of its severity. It is interesting to study F's dependence on the parameters of the system. The case of interest is where $I_0/S_0 \ll 1$, because this is the usual situation. The solution for F is described in Figure 4.2. If $T \gg 1$, then

Figure 4.2 Final size ($F = S_\infty/S_0$) as a function of the parameter T ($T = aS_0/(1-\alpha)$). Here $I_0/S_0 \ll 1$.

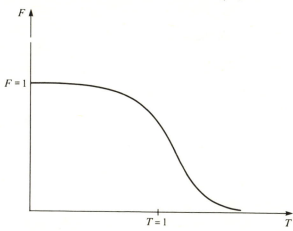

$F \sim 0$ and a large epidemic has occurred. But if $T \sim 1$, then $F \sim 1$, and only a few susceptibles are exposed. As T passes through the critical value $T = 1$, there is a dramatic change (referred to as a bifurcation) in the epidemic's final size. This result is a basic one in epidemiology. Note that for any fixed value of T, $S_\infty > 0$, so some susceptible will always survive the epidemic.

B. *Global bifurcation phenomena. Schistosomiasis: the use of thresholds in control of disease*

The Kermack–McKendrick threshold result is closely related to interesting phenomena that occur in many physical and biological systems. These are bifurcation phenomena, and we illustrate them with another epidemic model.

Schistosomiasis is a disease caused by a wormlike parasite (helminth). Mild forms are familiar to the United States; for example, swimmer's itch and clam digger's itch occur commonly in lakes and estuaries. More severe forms can be debilitating, and these forms are widespread in tropical regions around the world. In fact, this disease is second only to malaria in its global distribution.

Male and female helminths must mate in a host (human beings, swine, ducks, etc.). Thereafter, some of the fertilized eggs leave the host in its feces. The other eggs reach various organs in the host and cause degenerative diseases.

When an egg comes in contact with fresh water, it hatches as a larvae, called a miracidium. Next, the miracidia must find one of various species of snail to penetrate. Once a snail is infected, a large number of larvae, called circariae, are produced. The circariae swim freely; some forms can penetrate the skin of a host, others must be ingested. In the host, the circariae become helminths, and the cycle can continue.

This disease has features that separate it from most others. (a) The pathogen has two sexes, therefore requiring sexual reproduction somewhere in the disease cycle. (b) Superinfection is the rule, and the unit of infection is the number of parasites harbored rather than the number of infected hosts. (c) An intermediate host is required to complete the cycle outside the primary host.

A model for this disease has been formulated (MacDonald, 1965;

Nasell and Hirsch, 1973; Ludwig and Haycock, 1976). The following is a slight variation of their models. Let

H_n = mean number of worms infecting each host at time n

I_n = number of infected snails

δ = death rate of infected snails

μ = death rate of helminths in a host

c = number of helminths becoming established in a host due to each infected snail

S = total number of snails (assumed to be constant)

$P(H)$ = mean number of paired worms per infected host

Then

$$H_{n+1} = (1 - \mu)H_n + cI_n \qquad \text{(schistosomiasis}$$
$$I_{n+1} = (1 - \delta)I_n + bP(H_n)(S - I_n) \qquad \text{model)}$$

Calculation of P(H). First, we must evaluate $P(H)$. Suppose that the number of males and the number of females in a host have a Poisson distribution with parameter $H/2$: Thus,

$$\Pr[\text{number of males} = k] = e^{-H/2}(H/2)^k/k!$$
$$\Pr[\text{number of females} = l] = e^{-H/2}(H/2)^l/l!$$

and the joint distribution is

$$\Pr[\text{number of males} = k \text{ and number of females} = l]$$
$$= e^{-H}(H/2)^{k+l}/k!l!$$

We suppose that the number of pairs formed by k males and l females is $\min(k, l)$, so the number of paired helminths is $2 \min(k, l)$. Therefore,

$$P(H) = \sum_{k=0}^{\infty} \sum_{l=0}^{\infty} 2 \min(k, l)e^{-H}(H/2)^{k+l}/k!l!$$

It follows that

$$P(H) = H[1 - e^{-H}(I_0(H) + I_1(H))]$$

where I_0 and I_1 are modified Bessel functions, although we will not need this fact for the remainder of this section.

The static states of the schistosomiasis model give the endemic states of the disease. They are determined by the equations

$$H = (1 - \mu)H + cI$$
$$I = (1 - \delta)I + bP(H)(S - I)$$

Since we are studying the dependence of endemic states on the model's parameters, it is *very* important to nondimensionalize the problem. This ensures that changing one parameter will not unexpectedly change other model parameters because of hidden scale changes. Thus, we define

$$u = \frac{cI}{\mu I_0}, \qquad v = H/I_0$$

where I_0 = initial number of infected snails. The endemic equations become

$$u - v = 0 \qquad\qquad\qquad (4)$$
$$(\alpha + \beta P(v))u - P(v) = 0$$

where

$$\alpha = \frac{\mu \delta I_0}{cbS}, \qquad \beta = \frac{\mu I_0}{Sc}$$

From the first equation, we have $u = v$. Therefore, the problem reduces to solving

$$P(v) = \frac{\alpha v}{1 - \beta v}$$

Figure 4.3 portrays two typical parameter sets (α, β). The intersections of the curves correspond to solutions of the equation. Figure 4.4 depicts these solutions for α fixed. As β decreases through β^*, two new solutions appear. The upper branch is stable, the middle one unstable, and $v = 0$ is stable. That is, if v_0 exceeds v^*, v_n approaches the endemic states described by the upper branch. If $v_0 < v^*$, then $v_n \to 0$. β^* is a bifurcation point for β.

Now, suppose that the disease is endemic (so that $\beta < \beta^*$ and v

Figure 4.3. Static states of the schistosomiasis model. Case I: $V = 0$ is the only state. Case II: $V = 0$, $V = H_U$, and $V = H_S$ are static states; $V = 0$ and $V = H_S$ are stable; $V = H_U$ is unstable.

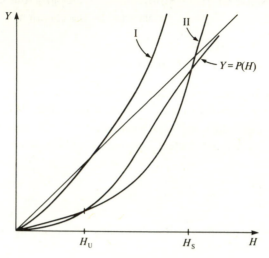

Figure 4.4. Description of static states for schistosomiasis and their stability as functions of β. Here α is fixed.

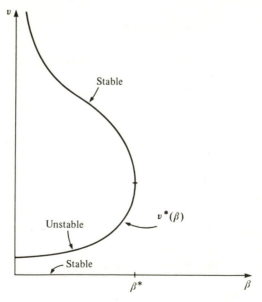

lies on the upper branch). If infected human beings are treated (μ increased) and the population is protected from contaminated water (c decreased) in such a way that

$$\beta \equiv \frac{\mu I_0}{Sc} > \beta^*$$

the disease will die out. These health measures can then be relaxed somewhat without an outbreak of disease.

C. *Discussion*

The static-state analyses in this section were successful because the equations for F in the threshold theorem and for (u, v) in (4) are simple enough to allow complete analysis. Introducing more realism to the models usually raises the level of difficulty in analyzing them. The study of the appearance of multiple static states is referred to as bifurcation theory. Most of this work rests on the implicit function theorem and on the Liapunov–Schmidt theory (see Vainberg and Trenogin, 1974).

Exercises

4.1 *(DeFinetti diagram).* This figure shows an equilateral triangle of altitude 1. Points P in this triangle are uniquely determined by their triangular coordinates (a, b, c).

Figure 4.5

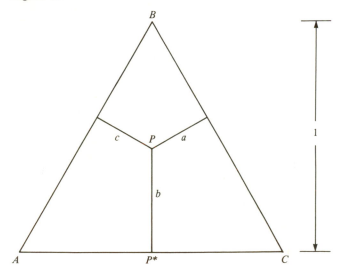

(a) Show that $a + b + c = 1$.
(b) Show that the distance from P^*, the projection of P onto the side AC, to C is $(2a + b)/\sqrt{3}$.

Consider a one-locus two-allele genetic trait in a population. The proportions of the population having genotypes A_1A_1, A_1A_2, and A_2A_2 are denoted, respectively, by D, $2H$, and R. We identify the vertex A with A_1A_1, B with A_1A_2, and C with A_2A_2. Then the point P having coordinates $(D, 2H, R)$ characterizes the population (at this locus). This is called a DiFinetti diagram.

(c) *(Hardy–Weinberg equilibrium)*. The genotype frequencies are in proportion $g^2 : 2g(1-g) : (1-g)^2$. Plot the points $(g^2, 2g(1-g), (1-g)^2)$ for $0 \le g \le 1$.

(d) Using a DiFinetti diagram, describe the dynamics of the genotype frequencies for the Fisher–Wright–Haldane model (1) when all fitnesses remain constant (as in the four cases in Figure 4.1).

4.2 *(Overlapping generations)*. One way to account for overlapping generations is to reformulate the basic model using continuous functions to describe the genotype frequencies. Now, let $D(t)$ = number of AA genotypes at time t, $2H(t)$ = number of Aa genotypes at time t, and $R(t)$ = number of aa genotypes at time t, where D, H, and R are now taken to be smooth functions of t. Suppose that these subpopulations reproduce according to Malthus's rule, with common birth rates b and death rates d_1, d_2, and d_3, respectively.

(a) Derive a system of ordinary differential equations for D, H, R, and $P(t)$ the total population size at time t.

(b) From this calculation, derive a system of differential equations for the genotype frequencies $x(t) = D(t)/P(t)$, $2y(t) = 2H(t)/P(t)$, $z(t) = R(t)/P(t)$, and the A-gene frequency $p(t) = x(t) + y(t)$.

(c) Consider slow selection by death: $d_i = d + \epsilon\Delta_i$, $i = 1, 2, 3$. Use the method of matched asymptotic expansions to analyze this system for $\epsilon \ll 1$.

4.3 *(van der Pol's equation)*. We consider a predator (Z) and prey (V), which interact in the following way. When V is large, Z increases, eventually causing the collapse of V to a low level. At the low level, the prey increase in numbers and the predators decrease. The prey population eventually explodes to high level, and the process repeats itself. In the absence of the predation Z, the prey population has a stable low level, perhaps controlled by competition, a stable high level, and an unstable intermediate equilibrium. A model for this can be derived by considering the deviation of prey and predators from the unstable equilibrium, say x for prey and u for predators. For example, a simple model describing these interactions is

$$dx/dt = \mu(x - x^3/3) - u$$
$$du/dt = x$$

where μ is a parameter representing the ratio of prey growth rate to predator response. The equilibrium curve $dx/dt = 0$ is a cubic, as depicted in the diagram. μ is taken to be small, $0 < \mu \ll 1$ here.

Figure 4.6

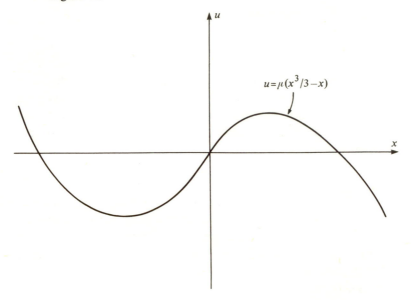

$$u = \mu(x^3/3 - x)$$

(a) Convert this problem to polar coordinates by introducing two new variables r (radius) and θ (angle) defined by $x = r \cos \theta$, $u = r \sin \theta$.

(b) Average the right-hand side of the equations for r and θ by integrating over θ from zero to to 2π while keeping r fixed. Describe the solution of these equations. It can be shown that this approximation accurately describes [to within order $O(\mu)$] the amplitude of the full (unaveraged) system. Using this fact, describe the behavior of $x(t)$, $u(t)$ as t increases.

5

Dispersal processes

The distribution of organisms in a geographical region, their movement, and the movement of nutrients cause interesting population phenomena. Numerous mathematical methods are available to study these processes. We first consider two examples of discrete-state discrete-site processes.

5.1   **Integer dispersal**
A geographical region can be lumped into discrete collection of sites. We visualize these as points on a plane having integer coordinates. The population numbers at these sites can be recorded and denoted by S_{ij}^n at the nth census of address (i, j). Less precise data than population sizes can be kept; for example, S_{ij}^n might be $+1$ if the site is occupied and 0 if the site (i, j) is vacant. Dispersal phenomena occur when contiguous sites can interact. Models will be specified by rules that tell how these states change with time. We assume that all changes according to the rules occur simultaneously at each tick of some clock.

A. *The game of life*
A site can be either occupied ($S = 1$) or vacant ($S = 0$). The rules for change are:
   (a) An occupied site having either two or three neighboring sites occupied will continue to be occupied (there are eight neighbors of each site).
   (b) Each occupied site having no, one, four, or more occupied neighbors becomes vacant.
   (c) Each vacant site becomes occupied if it has exactly three occupied neighbors.
This model was derived by Conway (1970). A great deal of work has gone into discovering what occupancy patterns persist as stable

92

configurations, which recur periodically, and which die out. Some examples are given in Figure 5.1.

B. *Spread of defoliating insects*

Now, we visualize each site as being a clump of vegetation. Each clump is either green ($S = 0$), infested with insects ($S = +1$), or defoliated ($S = -1$). Changes are governed by the following rules:

   (a) An infested site becomes defoliated.

   (b) A defoliated site becomes green.

   (c) A green site becomes infested if any of its four nearest neighbors is infested.

These rules are illustrated in Figure 5.2. In this, we see a wave of infestation propagating out from a single infested site (Figure 5.2*a*). Some initial configurations have quite interesting behavior. For example,

$$\begin{pmatrix} 0 & +1 \\ 0 & -1 \end{pmatrix}$$

acts as a pacemaker because it reproduces itself every four steps. This is shown in Figure 5.2*b*. A variety of propagation phenomena are possible in this model.

This model is easy to implement on a computer; however, it has not helped much in answering many questions about the system. For

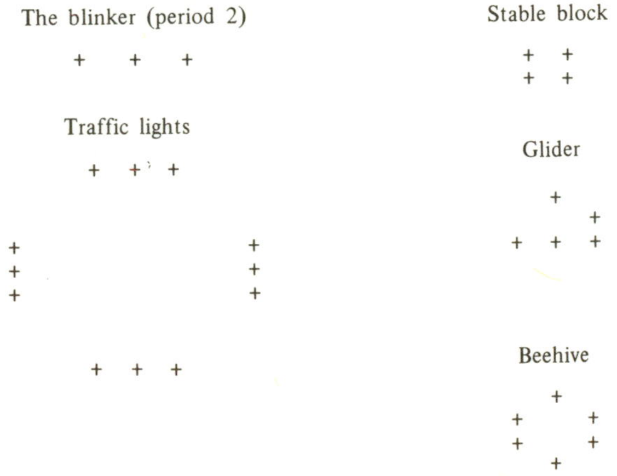

Figure 5.1. Some examples from the Game of Life.

The blinker (period 2)

   +   +   +

Stable block

   + +
   + +

Traffic lights

   + + +

Glider

     +
       +
   + + +

+   +
+   +
+   +

   + + +

Beehive

     +
  +   +
  +   +
     +

example, questions of interest are: Given an initial configuration, what will it evolve into? Given a developed pattern, what were the initial data? What are the self-reproducing units that act as pacemakers or focii of infestation? These are difficult questions to answer.

A precise formulation of this model is given by the formula

$$S_{ij}^{n+1} = \max(S_{i+1,j}, S_{i-1,j}, S_{i,j-1}, S_{i,j+1})_+ (1 - (S_{i,j}^n)^2)$$
$$- \left(\frac{S_{i,j}^n + 1}{2} \right) S_{i,j}^n$$

Models of this kind have been derived by Wiener and Rosenbleuth

Figure 5.2. Spread of defoliating insects, illustrated by the evolution of three infestation configurations. (*a*) A single wave propagates from the infestation site. (*b*) The initial block repeats itself every four steps, generating a sequence of waves propagating outward. (*c*) A spiral wave evolves.

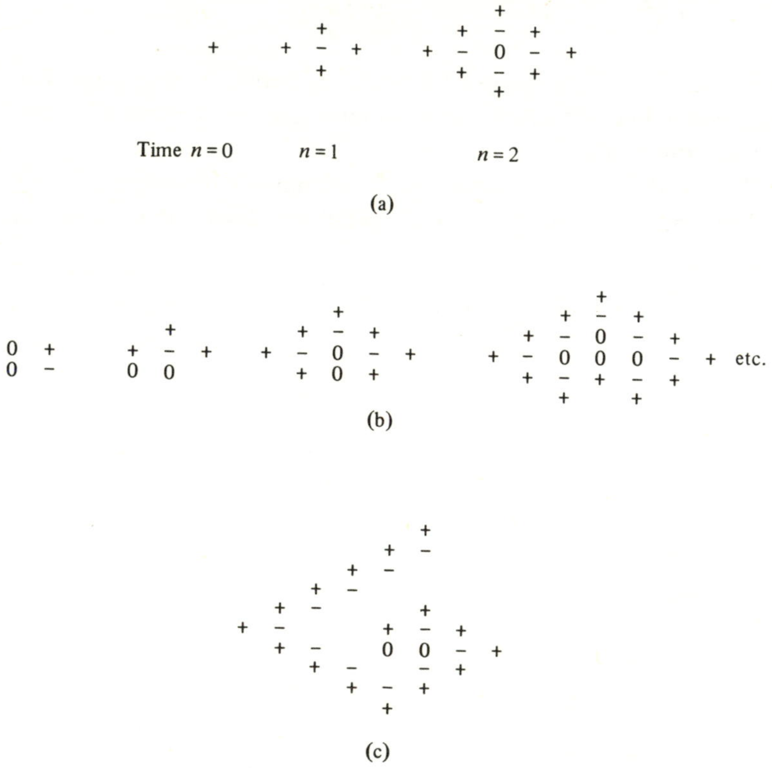

(a)

(b)

(c)

(1946), Moe et al. (1964), and Greenberg et al. (1978) to describe the dynamics excitable media such as heart muscle.

5.2 Diffusion approximations

A. *Random walk*

Consider contiguous sites in a one-dimensional habitat labeled $\ldots, m - 1, m, m + 1, \ldots$, and let $u_{m,n}$ denote the number of organisms at site m at time n. Suppose that

$$u_{m,n+1} = \frac{\kappa}{2} u_{m+1,n} + (1 - \kappa) u_{m,n} + \frac{\kappa}{2} u_{m-1,n}$$

where κ is a constant, $0 < \kappa < 1$. This assumption is equivalent to the random walk model, where the individuals at site m have probability $\kappa/2$ of moving left, $\kappa/2$ of moving right, and $1 - \kappa$ of remaining. The associated (Markov) transition matrix would be tridiagonal, with $1 - \kappa$ on the diagonal and $\kappa/2$ on the off diagonals.

If there are many sites and the time steps are not large, we might introduce a smooth function to describe $u_{m,n}$. Let $u(x, t)$ be such that

$$u_{m,n} = u(m \, \Delta x, n \, \Delta t)$$

where Δx and Δt are some small numbers. Substituting in the equation gives

$$u(m \, \Delta x, (n + 1) \, \Delta t) - u(m \, \Delta x, n \, \Delta t) = \frac{\kappa}{2}[u((m + 1) \, \Delta x, n \, \Delta t)$$
$$- 2u(m \, \Delta x, n \, \Delta t) + u((m - 1) \, \Delta x, n \, \Delta t)]$$

Next, keeping $x = m \, \Delta x$ and $t = n \, \Delta t$ fixed and expanding u in powers of Δx and Δt, we obtain

$$u(x, t + \Delta t) = u(x, t) + \frac{\partial u}{\partial t}(x, t) \, \Delta t + O(\Delta t^2)$$

$$u(x + \Delta x, t) = u(x, t) + \frac{\partial u}{\partial x}(x, t) \, \Delta x + \frac{1}{2} \frac{\partial^2 u}{\partial x^2}(x, t) \, \Delta x^2$$
$$+ O(\Delta x^3)$$

$$u(x - \Delta x, t) = u(x, t) - \frac{\partial u}{\partial x}(x, t) \, \Delta x + \frac{1}{2} \frac{\partial^2 u}{\partial x^2}(x, t) \, \Delta x^2$$
$$+ O(\Delta x^3)$$

Therefore,

$$\frac{\partial u}{\partial t}(x, t)\, \Delta t + O(\Delta t^2) = \frac{\kappa}{2}\frac{\partial^2 u}{\partial x^2}(x, t)\, \Delta x^2 + O(\Delta x^3)$$

If Δx, $\Delta t \ll 1$, but $\Delta x^2/\Delta t \sim 1$, then

$$\frac{\partial u}{\partial t} \sim \frac{\kappa}{2}\frac{\partial^2 u}{\partial x^2}$$

Therefore, we consider the equation

$$\frac{\partial u}{\partial t} = \frac{\kappa}{2}\frac{\partial^2 u}{\partial x^2} \qquad \text{(diffusion equation)}$$

as being an approximation to the original difference equation. A great deal of work has been invested in studying the solutions of this equation, notably by Fourier (1822) and many others since.

B. *Diffusion approximations to Markov chains*

The Fisher–Wright genetics model (Section 3.2B) with selection and mutation is difficult to analyze directly. There is a powerful method that gets around these obstacles, and it also involves approximation of discrete-time processes by continuous-time processes.

Markov chain models were stated in terms of a vector $\mathbf{p}_n$, which gives the distribution of the gene pool at time n, and the transition equation $\mathbf{p}_{n+1} = \mathbf{p}_n P$ involving the transition probability matrix P. This equation gives the forward evolution of the distribution vector $\mathbf{p}_n$. Here $\mathbf{p}_n$ will be approximated by a smooth function $\phi(x, t)$, where

$$(\mathbf{p}_n)_m = \phi\left(\frac{n}{2N}, \frac{m}{2N}\right)$$

denotes the mth component of $\mathbf{p}_n$. Then

$$(\mathbf{p}_{n+1})_m - (\mathbf{p}_n)_m = \phi\left(\frac{n+1}{2N}, \frac{m}{2N}\right) - \phi\left(\frac{n}{2N}, \frac{m}{2N}\right)$$

$$= \sum_k \phi\left(\frac{n}{2N}, \frac{k}{2N}\right)(P_{k,m} - \delta_{k,m})$$

As $N \to \infty$, this equation passes to the form

$$\frac{\partial \phi}{\partial t} = \frac{1}{2}\frac{\partial^2}{\partial x^2}[v(x)\phi] - \frac{\partial}{\partial x}[m(x)\,\phi]$$

provided that certain additional conditions are satisfied.

The motivation for this comes from the *central limit theorem*, which states that *any* random variable, discrete or continuous, is in a definite sense approximated by a normally distributed random variable. A normal random variable is characterized by its mean and its variance (say μ and $t\sigma^2$, respectively), and the probability density function

$$u(x, t) = (1/\sqrt{2\pi t\sigma^2}) \exp\left[(-\tfrac{1}{2}) \left(\frac{x - \mu}{\sigma\sqrt{t}} \right)^2 \right]$$

satisfies the diffusion equation

$$\frac{\partial u}{\partial t} = \frac{\sigma^2}{2} \frac{\partial^2 u}{\partial x^2}$$

This approximation is developed and these ideas applied to the Fisher–Wright models in the Appendix.

5.3 Transform methods: linear stability theory for diffusion equations

A. *Plankton blooms*

Let us consider a population of phytoplankton that is distributed over a one-dimensional habitat. Say they occupy an interval $-L \leq x \leq L$, and that plankton meeting the boundary are eliminated (e.g., by unfavorable habitat, predators, etc.). At each point, we take a Malthus reproduction model with intrinsic growth rate α. The model of this is

$$\frac{\partial u}{\partial t} = \frac{\sigma^2}{2} \frac{\partial^2 u}{\partial x^2} + \alpha u, \qquad -L < x < L$$
$$u(0, t) = u(L, t) = 0$$

where $u(x, t)$ is the density of phytoplankton at position x.

Let us ignore at first the boundary conditions, and analyze solutions of the equation alone. We look for solutions in the form $u = e^{ikx + pt}$. This amounts to taking a Fourier transform in x and a Laplace transform in t. Here k is called the *wave number*, and p is the *amplification rate*.

If such a function is to be a solution, p and k must satisfy the condition

$$p = -(\sigma^2 k^2/2) + \alpha \qquad \text{(dispersion relation)}$$

which is known as the *dispersion relation*. Attention is restricted to values of k, σ, and α for which $p > 0$, because it is for these values that solutions can grow. We plot the curve $p = 0$. For values of α and k for which $p > 0$, $e^{pt + ikx}$ grows. These terms will grow out of the regime where the Malthus model is reasonable, but they give an indication of what spatial structure will be observed. These solutions are called unstable modes (see Figure 5.3).

The boundary conditions select only certain values of k that can be used in the solution. In particular,

$$u(x, t) = \sum_{n=1}^{\infty} A_n \exp(p_n t) \sin(n\pi x/L)$$

So the appropriate wave numbers are $k_n = \pm n\pi/L$ and the amplification rates are $p_n = \alpha - (k_n^2 \sigma^2/2)$. Now, for fixed α there are at most finitely many unstable modes, as shown in Figure 5.4.

Figure 5.3. Neutral stability curve for the plankton bloom model ($p = 0$) as a function of the wave number k.

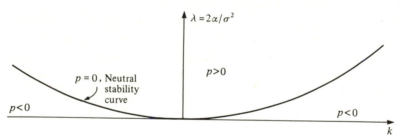

Figure 5.4. Dashed line depicts $\lambda =$ constant. The appropriate wave numbers in this case lie on it. Those within the curve are for unstable modes; those outside are stable.

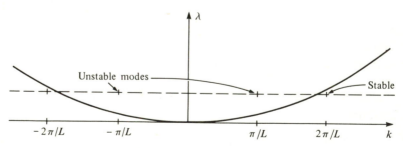

The nth mode is unstable if

$$(n\pi/L)^2 \, (\sigma^2/2\alpha) < 1$$

The first mode is the most unstable because it has the largest amplification rate. The instability condition for the most unstable mode is

$$(\pi/L)^2(\sigma^2/2\alpha) < 1$$

Thus, we see that there is a critical length L for which a bloom can develop: $L^* = \pi\sigma/(2\alpha)^{1/2}$. When $L < L^*$, all modes are decaying and the population dies out. However, if $L > L^*$, there are growing modes, so the plankton bloom. Its profile will have the shape of $\sin(\pi/L)x$.

The values of σ and α are fixed by the organism and habitat. If the population is restricted to too small a region, it will die out, but if its habitat exceeds a critical size (L^*), a bloom develops. This suggests that plankton patches have a certain minimum patch size.

B. *Diffusive instability*

An interesting phenomenon can occur between interacting species that diffuse at differing rates. This will be illustrated here with an artifical model of two interacting species:

$$\frac{\partial u}{\partial t} = \mu \frac{\partial^2 u}{\partial x^2} + au + bv$$
$$\frac{\partial v}{\partial t} = \nu \frac{\partial^2 v}{\partial x^2} + cu + dv$$

where μ, ν (the diffusivities) and a, b, c, d (interaction rates) are fixed constants. Although a model of this kind is unrealistic by itself, one of this form will result whenever the stability of a uniform static state of a nonlinear diffusion system is to be tested.

Fourier theory guarantees that the solution of this system can be found as a sum of terms having the form $\exp(ikx + pt)$. Substituting $u, v \sim e^{ikx+pt}$, we have

$$pI = \begin{pmatrix} a - \mu k^2 & b \\ c & d - \nu k^2 \end{pmatrix}$$

so the p-values are the eigenvalues of the matrix

$$B = \begin{pmatrix} a - \mu k^2 & b \\ c & d - \nu k^2 \end{pmatrix}$$

The phenomenon of interest here arises when the interaction matrix

$$A = \begin{pmatrix} a & b \\ c & d \end{pmatrix}$$

is stable (all its eigenvalues have negative real parts) but B is unstable for some k.

The following example shows that this is possible. Let

$$A = \begin{pmatrix} 1 & 1 \\ -100 & -10 \end{pmatrix}$$

its eigenvalues are $-(9/2) \pm (i/2) \sqrt{279}$. The characteristic equation for B is

$$\lambda^2 - (9 + (\mu + \nu)k^2)\lambda + 90 + k^2(10\mu - \nu) + \mu\nu k^4 = 0$$

The discriminant of this quadratic is

$$\left(\frac{9 + (\mu + \nu)k^2}{2} \right)^2 - (90 + k^2(10\mu - \nu) + \nu\mu k^4)$$

If the second term is positive, the largest eigenvalue of B has positive real part. This happens if we take $\mu = 1/\nu$ with ν sufficiently large.

Thus, the diffusion terms excite the instability. These phenomena are called *diffusive instabilities,* and they arise in a wide variety of biological and physical phenomena. The following theorem gives conditions that rule out diffusive instability:

Theorem. Let u be a vector having N components. Suppose that

$$\frac{\partial u}{\partial t} = \mathcal{D} \frac{\partial^2 u}{\partial x^2} + Au$$

where $\mathcal{D}$ is a diagonal matrix having nonnegative components and A is an $N \times N$ matrix. Then if $\frac{1}{2}(A^{\mathrm{Tr}} + A)$ is stable, there can be no diffusive instability.

Proof. Let $(u \cdot v) = \int u(x) \cdot v(x) \, dx$. Since $\partial(u \cdot u)/\partial t \leq -K(u \cdot u)$ for some positive constant K, we see that $(u \cdot u) \to 0$ as $t \to \infty$.

5.4   Pattern formation

Regular patterns of organisms occur frequently in a variety of biological settings, and they are caused by a variety of mechanisms. Two examples are presented here, which illustrate two different mechanisms for pattern formation.

The first describes the patterned growth of bacteria in response to a diffusing nutrient. The bacteria do not move; instead, they act as a living record of past events that have been favorable or unfavorable to their growth. The key ingredients are the nutrient, which promotes growth, and the acid, which inhibits growth. When the process is complete, there is no detectable pattern structure for either the nutrient or the acid; only the differing cell population numbers at various points indicate the way in which nutrient and acid approached their uniform distributions.

The second example illustrates a reverse mechanism. This shows how a population's genetic structure can adapt to a patterned structure of the environment that influences genetic fitness. In this case, the population adapts to a patterned profile of fitness coefficients.

A. *Dynamic patterns: the crystal test*

Bacteria can grow in spatial patterns in response to diffusion of a needed nutrient. Experiments that demonstrate this are presented here. The model describes an immobile bacterial population distributed on a petri dish. The diffusing nutrient is taken up by growing cells, and acid is produced as a by-product of cell growth. Numerical computations show that the model exhibits behavior similar to that observed in experiments.

A novel feature of the model is that the cell's growth has a hysteretic dependence on the amount of nutrient and acid present as shown in Section 1.3B. It is known that hysteretic kinetics can lead to spatial patterns in chemical systems, such as Liesegang rings formed by precipitating colloids. In that case, the hysteresis is described by Ostwald's theory of supersaturation. Other hysteretic systems that lead to patterns arise in electrophysiology and epidemiology.

Experiment. A lawn of bacteria is fixed on a circular agar gel that contains an acid buffer and minimal growth chemicals, except that these cells need the amino acid histidine to grow (these are called his-

tidine auxotrophs). A drop of histidine solution is placed in the center
of the gel, and it diffuses out. The cells grow in response to the histidine in a pattern of concentric circles as shown in Figure 5.5. This is
a standard procedure, called the *crystal test,* for determining cell
characteristics.

Crystal test model. The bacterial population size at radius r ($0 \leq r$
$\leq R$) on the plate is denoted by $B(r, t)$, the concentration of histidine
by $H(r, t)$, and the pH by the buffer concentration $G(r, t)$. These
functions are determined by the equations

$$\frac{dB}{dt} = \alpha VB \qquad \text{(cell growth)}$$

$$\frac{\partial H}{\partial t} = D\,\Delta H - \beta VB \qquad \text{(histidine diffusion and uptake)}$$

$$\frac{\partial G}{\partial t} = D'\,\Delta G - \gamma VB \qquad \begin{array}{l}\text{(buffer diffusion and cell acid}\\ \text{production)}\end{array}$$

Diffusion is described by the operator Δ in radial coordinates

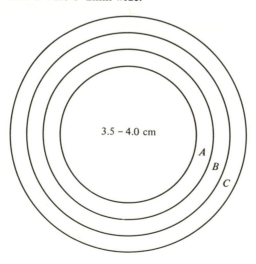

Figure 5.5. Rings of concentric growth of bacteria. Histidine
auxotrophs are plated in top layer agar on which 2 µl of a 1
molar histidine solution was placed at the center. The regions *A,*
B, and *C* are regions of no bacterial growth: $A \sim$ 5mm, $B \sim$
4mm, and $C \sim$ 2mm. The bands of growth that separated *A, B,*
and *C* were 1–2mm wide.

$$\Delta u = \frac{1}{r} \frac{\partial}{\partial r} r \frac{\partial}{\partial r} u$$

The constants D and D' are the diffusivities of H and G, respectively. The kinetic terms involve the function V, which describes the growth competence of sites in the dish; αV is the bacterial growth rate, βV the uptake rate of histidine, and γV the acid production by growing cells. The function V depends on the amounts of histidine and buffer. It could be taken to have the form

$$V = \frac{H}{K + H} \frac{G}{K' + G}$$

where K and K' are Michaelis (or saturation) constants. However, it is known from experiments that for adequate histidine concentrations, cells have a hysteresis in growth depending on pH. When growing, cells will continue to grow even as pH decreases until a threshold is reached at which growth stops. As pH is increased from this threshold, growth does not begin again until a higher, more favorable pH threshold is reached. This is described in Figure 5.6.

The situation is somewhat more complicated by similar thresholds for histidine. This is illustrated in the GH-phase plane in Figure 5.7,

Figure 5.6. Cells grow when $V > 0$. As pH decreases in the presence of adequate histidine, V drops to zero quickly. As pH increases, V remains 0 until the threshold $\hat{T}_{on}$ is reached, at which point V quickly moves to the upper branch.

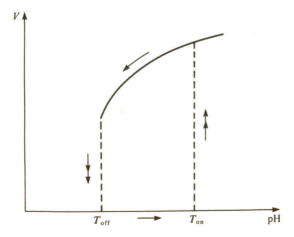

where ϕ denotes the threshold at which cell growth stops and ψ denotes that at which cells again begin to grow. Figure 5.8 shows the dependence of histidine uptake on G as G passes into the overlap region.

The results described here depend on choices for the initial data and the thresholds ϕ and ψ. A variety of different patterns are possible for various choices of them. The exact form of the functions ϕ and ψ is unknown, but work is under way to determine them experimentally.

Figure 5.7. GH-phase plane provides a useful description of the model dynamics. Here are typical choices for the turn-off threshold (ψ) and the turn-on threshold (ψ). The overlapping region corresponds to the region between T_{off} and T_{on} in Figure 5.6.

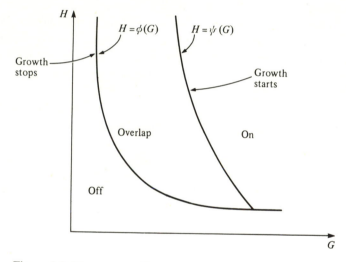

Figure 5.8. These curves illustrate how histidine uptake varies with G.

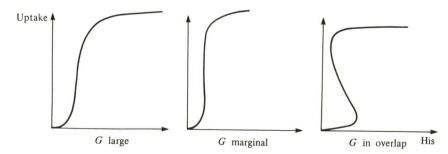

Uniform initial distributions of bacteria and buffer are given by

$$B(r, 0) = B^0 \quad \text{and} \quad G(r, 0) = G^0$$

and the initial histidine distribution is given by $H(r, 0) = H^0(r)$, as depicted in Figure 5.9.

Analysis of the model. Experimental values of the parameters indicate that diffusion is dominated by kinetics ($D, D' \ll \beta, \gamma$) when $V > 0$. This suggests the following heuristic arguments, which we describe in the GH-phase plane. Figure 5.10a shows the initial data and the thresholds. Figure 5.10b shows the result of kinetics acting on these data. Note that if we ignore diffusion, then $dH/dG = \beta/\gamma$. Thus, the initial data move toward the curve $H = \phi(G)$ along straight lines. Next, diffusion acts on this distribution, slowly raising portions toward the threshold $H = \psi(G)$. Figure 5.10c shows the results at the point where certain cells are again turned on. These drive the turned-on values (G, H) toward the curve $H = \phi(G)$. This process continues to give repeated switching on and off of various sites. It stops when there no longer remains enough histidine or buffer to again reach $H = \psi(G)$. The result is a spatial pattern of growth. Diffusion will eventually redistribute the remaining histidine and acid uniformly.

Approximate behavior of the model can be deduced by iterating the kinetic and diffusion phases. The kinetics were described in the preceding paragraph. Motion is along straight lines having slope β/γ in

Figure 5.9. Initial histidine distribution.

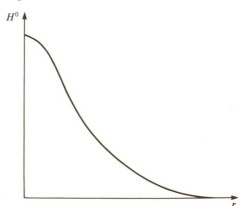

the *GH*-plane. Once all growth stops, the model reduces to two diffusion equations for *G* and *H*. These can be solved in terms of Fourier–Bessel expansions, and these, in turn, used to approximate where the next competent sites will be. This procedure and the construction of appropriate boundary layers for matching these solutions together is lengthy and will not be presented here.

Figure 5.10. *GH*-phase plane. (*a*) Initial distributions and thresholds. (*b*) Approximate distribution of *G* and *H* after initial kinetics. (*c*) Approximate distribution after diffusion has moved up to the turn-on threshold.

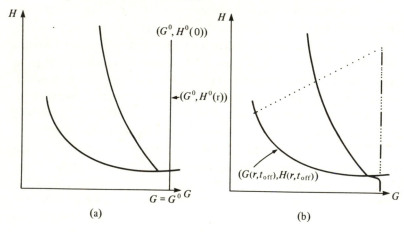

(a)

(b)

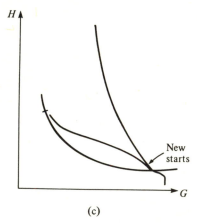

(c)

Numerical example. This section presents the results of a numerical simulation of the crystal test model that is based on a method of lines solver. The parameters chosen for the system are based on rescaling, nondimensionalizing, and substituting realistic values into the result. The choice of the threshold curves ϕ and ψ is difficult because there are few data available about this phenomenon. However, we define

Growth stops: $\{(G, H): \quad \min(G, H) = 1\}$

Growth starts: $\{(G, H): \quad H \geq 1, G = (a/H) + b\}$

These are shown in Figure 5.11.

The function V is taken to be 1 if a site is competent and zero otherwise. Figure 5.12 shows the result of the calculations for $\alpha = 1$, $\beta = 5$, $\gamma = 5$, $D = 1E - 3$, $D' = .5E - 3$, $G^0 = 500$, $a = 100$, $b = 100$, $H^0 = 2000$ if $r \leq R/10$, $= 0$ for $r > R/10$.

The numerical result described in Figure 5.12d corresponds to the passage of approximately 3 days for growing bacteria as observed in experiments. It compares favorably with the experimental observations in Figure 5.5. These results were reported in (Hoppensteadt and Jager, 1980).

Figure 5.11. *GH*-phase plane, depicting threshold curves used in the numerical computation.

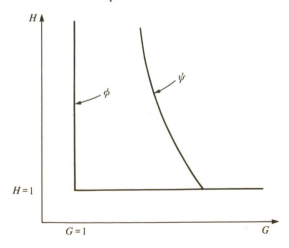

B. *Asymptotic (static) patterns: genetic clines*

The phenomenon of a genetic cline has been observed in a great many biological settings. A simple example of this phenomenon is given by green and brown praying mantises. We model color as a one-locus two-allele genetic trait. The green mantises will have a selective advantage in green habitats, but a selective disadvantage in a brown habitat because they are easier targets for predators, such as birds; similarly for brown mantises.

Now consider contiguous brown and green habitats. Given that green is genetically dominant, what will the genetic structure of the population be? It would seem that random movements of the mantises would maintain the green allele in the brown habitat and brown in the green habitat, at least near the boundary. Although the situation is

Figure 5.12. (*a*) and (*b*) Illustrations of the descriptions in Figure 5.10.

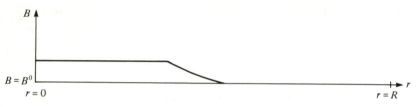

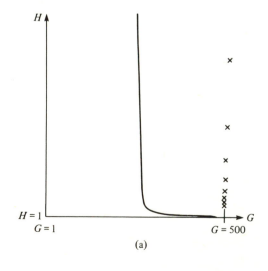

(a)

complicated by preferential mating and inhibitions against changing habitats, this system does give an indication of a mechanism for maintaining a gene in an unfavorable environment.

Fisher's equation (see Section 5.5A) can be used to study this phenomenon further. Let us consider a one-locus two-allele trait in a population distributed along an interval, say $-1 \le x \le 1$. The frequency of one allele in the gene pool will be assumed to be modeled by Fisher's equation with dispersal

$$\frac{\partial u}{\partial t} = \kappa \frac{\partial^2 u}{\partial x^2} + s(x)u(1 - u)((1 - h)u + h(1 - u))$$

where κ measures the movement of organisms, $s(x)$ is the selective

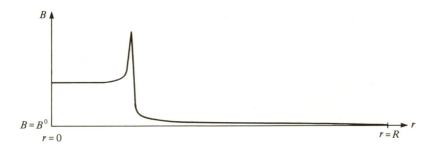

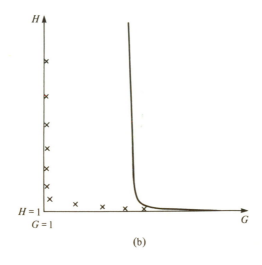

(b)

intensity at location *x,* and *h* is a measure of the relative selective advantage enjoyed by homozygotes for the *A* allele over heterozygotes.

We rewrite this equation as

$$\frac{\partial u}{\partial t} = \frac{\partial^2 u}{\partial x^2} + \lambda g(x)u(1 - u)((1 - h)u + (1 - u)h) \qquad (1)$$

and add boundary conditions corresponding to reflecting boundaries

$$\frac{\partial u}{\partial x}(+1, t) = \frac{\partial u}{\partial x}(-1, t) = 0 \qquad (2)$$

Now $\int_{-1}^{1} |g(x)| \, dx = 1$, λ measures the ratio of selective intensity to diffusivity, and $t = \kappa t'$.

Figure 5.12. (*c*) Illustration of the description in Figure 5.10.

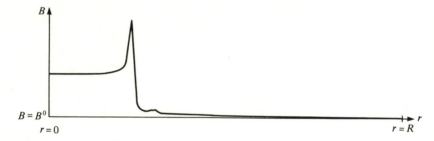

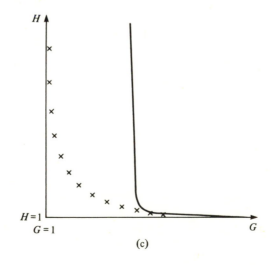

(c)

In the article by Fleming (1975), various possible static solutions to (1) and (2) were derived. These are of interest because they are candidates for describing the eventual state to which a given initial distribution will evolve. In particular, the following theorem was proved.

Theorem (Fleming). If g is a piecewise continuous function taking on both positive and negative values and satisfying

$$\int_{-1}^{1} g(x) \, dx < 0$$

there is a positive number $\lambda = \lambda_1$ such that the solution $u_0 \equiv 0$ of (1) and (2) is stable provided that $0 < \lambda < \lambda_1$. If $f''(0) < 0$, then for $\lambda > \lambda_1$, there is a static solution $u = u(x, \lambda)$ of (1) and (2) such that

Figure 5.12. (*d*) Distributions after a time comparable to that at which Figure 5.5 was drawn.

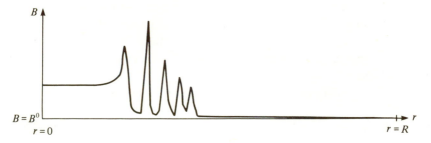

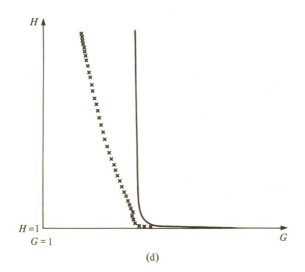

(d)

(a) $0 < u(x, \lambda) < 1$ for $-1 \le x \le 1$.
(b) $\max u(x, \lambda) \to 0$ as $\lambda \to \lambda_1 + 0$.
(c) $u_0 \equiv 0$ is unstable for $\lambda > \lambda_1$, whereas $u(x, \lambda)$ is stable for $0 < \lambda - \lambda_1 \ll 1$.

The condition on g requires that the "average fitness" of the allele be negative in the habitat: $\int_{-1}^{1} g(x)\, dx < 0$. However, since g takes on both positive and negative values, there are intervals where a is favored. In spite of this, the first result shows that a will be removed from the population everywhere provided that the parameter λ is smaller than $\lambda_1(\lambda < \lambda_1)$. On the other hand, as λ increases through the critical value $\lambda = \lambda_1$, a new stable state $u(x, \lambda)$ appears. Since $0 < u(x, \lambda) < 1$, this corresponds to a polymorphic state in the population where both alleles are manifested. Thus, if the selective intensity and the dispersal rate are correctly "tuned," a stable polymorphism will be maintained in the population.

It is of some interest to study the stability of this polymorphism and the evolution of various initial data to it. This can be done partly by using a perturbation scheme to construct the solutions of the model. This approach has been widely used in the study of many physical systems.

The stability results are summarized in the following theorem.

Theorem. Let $\epsilon = \lambda - \lambda_1$, and consider the problem for small positive values of ϵ. For any data of the form $\phi(x) = \epsilon U(x, \epsilon)$, where U is a smooth function satisfying the boundary conditions, the solution of the problem is given in the form

$$u(x, t, \epsilon) = \epsilon(C(\epsilon t)v_0(x) + W(x, t, \epsilon)) + O(\epsilon^2)$$

where $\max | W(x, t, \epsilon)| \le Ke^{-\delta t}$ for $t \ge 0$ and some positive constants K and δ, which are independent of t. The function $v_0(x)$ is the positive solution to the linear problem

$$\frac{d^2v_0}{dx^2} + \lambda_1 g(x)f'(0)v_0 = 0, \qquad \frac{dv_0}{dx}(1) = \frac{dv_0}{dx}(-1) = 0$$

which satisfies $\int_{-1}^{1} v_0^2(x)\, dx = 1$. [This normalization of v_0 differs from Fleming's, for his $v_0(x) = v_0(x)/v_0(-1)$.] The coefficient C is a function of the slow time variable $\sigma = \epsilon t$, and it is determined by the differential equation

$$dC/d\sigma = (1/\lambda_1)C + H_1C^2, \qquad C(0) = \int_{-1}^{1} U(x, 0)v_0(x) \, dx$$

where

$$H_1 = f''(0)(\lambda_1/2) \int_{-1}^{1} v_0^3(x)g(x) \, dx$$

Finally, the error term $O(\epsilon^2)$ holds uniformly for $0 \le t < \infty$ as $\epsilon \to 0$.

It is shown in Hoppensteadt (1975a) that the hypotheses of this theorem ensure that the perturbation result derived by Hoppensteadt and Gordon (1975) is applicable. In this way the theorem is established.

It is easily shown that for any positive integer n,

$$\int_{-1}^{1} v_0^n(x)g(x) \, dx > 0$$

Thus, the constant H_1 is negative. Moreover, if we write

$$U(x, 0) = av_0(x) + w(x)$$

where

$$\int_{-1}^{1} v_0(x)w(x) \, dx = 0$$

then $C(0) = a > 0$. Solving the equation for $C(\sigma)$, we have

$$C(\sigma) = \exp(\sigma/\lambda_1)C(0)$$
$$\left[1 - \frac{C(0)f''(0)}{2} \lambda_1^2 \int_{-1}^{1} v_0^3(x)g(x) \, dx(\exp(\sigma/\lambda_1) - 1) \right]$$

Combining this with the theorem, we see that

$$u(x, t, \epsilon) = \epsilon C(\epsilon t)v_0(x) + O(\epsilon e^{-\delta t}) + O(\epsilon^2)$$

As $\sigma \to \infty$, C approaches a limit:

$$C(\infty) = -2 / \left(f''(0)\lambda_1^2 \int_{-1}^{1} v_0^3(x)g(x) \, dx \right)$$

like the exponential $\exp(-\sigma/\lambda_1)$. Therefore, for $t \to \infty$, we have

$$u(x, \infty, \epsilon) = -\left(2\epsilon/f''(0)\lambda_1^2 \int_{-1}^{1} v_0^3(x')g(x')\,dx' \right)v_0(x) + O(\epsilon^2)$$

with $u(x, t, \epsilon) - u(x, \infty, \epsilon) = O(\epsilon e^{-\epsilon t/\lambda_1})$. This is an approximation [up to order $O(\epsilon^2)$] to the polymorphic state, and corresponds to the result that would have been obtained by Fleming if the Liapunov–Schmidt method had been used. Moreover, this state is approached at the rate $\exp(-\epsilon t/\lambda_1)$, which illustrates the evolution of u on a slow time scale.

5.5 Wave propagation: dynamic patterns

A. *Fisher's equation*

One of the first nonlinear diffusion equations to be analyzed was one proposed by R. A. Fisher (1936) to model the spatial distribution of an advantageous gene.

Consider a one-locus two-allele genetic trait in a large randomly mating population. Let $u(x, t)$ denote the frequency of one gene, say A, in the gene pool at x. If A is dominant, then the analog of the Fisher–Wright model for continuous-time approximation is

$$du/dt = s(x)u(1 - u)$$

where $s(x)$ measures the selective intensity at x. If different sites are connected by random migration of the organism, the gene pool can be described by a diffusion model:

$$\frac{\partial u}{\partial t} = \kappa \frac{\partial^2 u}{\partial x^2} + s(x)u(1 - u) \qquad \text{(Fisher's equation)}$$

Wave of advance of A ($s(x) \equiv s > 0$). Suppose that $u(x, 0) = 0$ for $|x| > 1$, and $u(x, 0) = 1$ for $|x| < 1$. Clearly, the A genes will diffuse to the right and left. One way to describe this is in terms of constant gene frequency. That is, we attempt to find $X(t)$ such that

$$\int_{|x| \geq X(t)} u(x, t)\,dx = C \ll 1$$

Consider the right-half line alone. Since u is small, we will ignore the quadratic terms and consider for $x \geq X(t)$ the equation

$$\frac{\partial u}{\partial t} = \kappa \frac{\partial^2 u}{\partial x^2} + su$$

Setting $J(y, t) = u(\sqrt{\kappa} y, t)e^{-st}$, we have $\partial J/\partial t = \partial^2 J/\partial y^2$. For $X \gg 1$, we expect that u can be approximated by the solution of this equation that satisfies the initial condition $J(y, 0) = \delta(y)J_0$. The solution is

$$J(y, t) = (J_0/\sqrt{4\pi t}) \exp(-y^2/4t)$$

Therefore,

$$u(x, t) = \frac{J_0}{\sqrt{4\pi t}} \exp\left(-\frac{x^2}{4\kappa t} + st\right)$$

It follows that

$$\frac{C}{2} = \int_X^\infty u \, dx$$

$$= \frac{J_0 e^{st}}{\sqrt{4\pi t}} \int_X^\infty \exp\left(-\frac{x^2}{4t\kappa}\right) dx \sim \frac{J_0}{\sqrt{4\pi t}} \exp\left(st - \frac{X^2}{4t\kappa}\right)$$

Therefore,

$$X(t) \sim 2t(s\kappa)^{1/2} + o(t) \quad \text{as} \quad t \to \infty$$

We deduce that the velocity of propagation will be approximately $2(s\kappa)^{1/2}$. This argument is a slight modification of Fisher's original one.

Steady progressing wave solutions. Kolmogoroff et al. (1936) analyzed Fisher's equation by looking for steady progressing waves. Let

$$u(x, t) = U(x + ct)$$

where U and c are to be determined. This form of solution is called a *steady progressing wave*, and $\xi = x + ct$ is the progressing wave variable.

As boundary conditions, we take

$$u(-\infty, t) = 0 \qquad u(+\infty, t) = 1$$

Substituting these into Fisher's equation gives

$$k \frac{d^2 U}{d\xi^2} = c \frac{dU}{d\xi} + sU(1 - U)$$

$$U(-\infty) = 0, \qquad U(+\infty) = 1$$

Let $V = dU/d\xi$. Then

$$dU/d\xi = V, \qquad dV/d\xi = (c/k)V - (s/k)U(1 - U)$$

The following analysis establishes the existence of solutions.

There are two rest points, $U = 0$, $V = 0$ and $U = 1$, $V = 0$. Linearizing about the first, we obtain as the coefficient matrix

$$\begin{pmatrix} 0 & 1 \\ -s/\kappa & c/\kappa \end{pmatrix}$$

whose eigenvalues are

$$\frac{c}{\kappa} \pm \left[\left(\frac{c}{\kappa} \right)^2 - \frac{4s}{\kappa} \right]^{1/2}$$

Linearizing about the second, we have the coefficient matrix

$$\begin{pmatrix} 0 & 1 \\ s/\kappa & c/\kappa \end{pmatrix}$$

whose eigenvalues are

$$\frac{c}{\kappa} \pm \left[\left(\frac{c}{\kappa} \right)^2 + \frac{4s}{\kappa} \right]^{1/2}$$

The rest point $(0, 0)$ must be unstable if the left boundary condition is to be satisfied. This requires that $c > 0$. If $c < 2(s\kappa)^{1/2}$, the origin is a spiral point, so any solution approaching it must eventually become negative. Thus, we must have $c \geq 2(s\kappa)^{1/2}$. The rest point $(1, 0)$ is a saddle point.

Next we prove:

Theorem (Kolmogoroff et al.). For any $c \geq 2(s\kappa)^{1/2}$, there is a progressing wave $u(x, t) = U_c(x + ct)$ that satisfies
  (a) $0 \leq U \leq 1$.
  (b) $U(-\infty) = 0$, $U(+\infty) = 1$.
  (c) $u(x, t)$ satisfies Fisher's equation.

Proof. For $c \geq c_{\min} = 2(\kappa s)^{1/2}$, there is a straight line l as shown in Figure 5.13.

Letting $\xi \to -\infty$ is equivalent to setting $\eta = -\xi$ and studying the solutions as $\eta \to +\infty$. The phase plane now becomes like Figure 5.14. Thus, the solution emanating from (0, 1) must approach the origin. This completes the proof.

The remarkable result here is that there is a continuum of waves all having different speeds. The waves appear as traveling fronts.

These waves can be expanded in terms of the reciprocal wave speed $1/c$ (Canosa, 1973; Hoppensteadt, 1975b). The result is that

$$u(x, t) = U_0 \left(\frac{2(s\kappa)^{1/2}}{c^2} (x + ct) \right)$$
$$+ \frac{2(s\kappa)^{1/2}}{c^2} U_1 \left(\frac{2(s\kappa)^{1/2}}{c^2} (x + ct) \right) + \dots$$

where

$$U_0(\xi) = 1/(1 + \exp(\xi))$$

and

$$U_1(\xi) = \frac{e^\xi}{(1 + e^\xi)^2} [4U_0(0) + \log(\xi e^\xi/(1 + e^\xi)^2)]$$

Figure 5.13. Phase portrait of the Kolmogoroff–Petrovsky–Picscounov system.

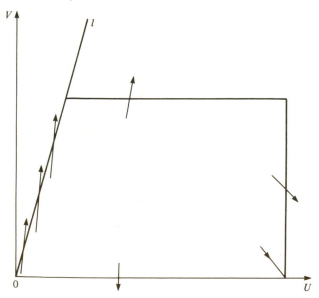

Surprisingly, this approximation gives three-place accuracy even when $c = c_{min}$.

Stability of steady progressing waves. Various investigations of the stability of the progressing wave solutions to Fisher's equation have been made. For example, Kolmogoroff et al. showed that the initial data

$$u(x, 0) = \begin{cases} 1 & \text{for} \quad x > 0 \\ 0 & \text{for} \quad x < 0 \end{cases}$$

evolve into a wave having the slowest speed of propagation, $c = c_{min}$. These results were extended by Hoppensteadt (1975b), by Aronson (1976), Weinberger, and by McKean. The analysis given here was derived in Hoppensteadt (1975b).

The question is formulated in the following way. Given $c > c_{min}$ and U_c what conditions on $u(x, 0)$ ensure that

$$\sup_{-\infty < x < \infty} |u(x, t) - U_c(x + ct)| \to 0$$

as $t \to \infty$?

Let $\tilde{\psi}(x, t) = u(x, t) - U_c(x + ct)$. This function satisfies

Figure 5.14. Phase portrait of the Kolmogoroff–Petrovsky–Picscounov system ($\eta = -\xi_1$).

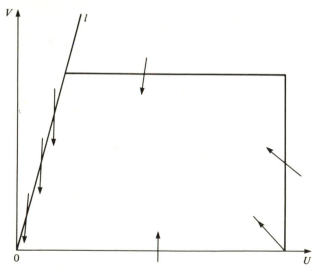

$$\frac{\partial \tilde{\psi}}{\partial t} = \frac{\partial^2 \tilde{\psi}}{\partial x^2} + s(\tilde{\psi} + U_c)(1 - \tilde{\psi} - U_c) - sU_c(1 - U_c)$$

(Here and below $\kappa = 1$.) Setting $z = x + ct$ and $\psi(z, t) = \tilde{\psi}(x, t)$, we obtain

$$\frac{\partial \psi}{\partial t} = \frac{\partial^2 \psi}{\partial z^2} + c\frac{\partial \psi}{\partial z} + s(1 - 2U_c(z))\psi + \mathcal{H}(\psi)$$

where $\mathcal{H}(\psi) = o(|\psi|)$ as $|\psi| \to 0$. We first estimate the solution of the linear part of the equation

$$\frac{\partial \hat{\psi}}{\partial t} = \frac{\partial^2 \hat{\psi}}{\partial z^2} + c\frac{\partial \hat{\psi}}{\partial z} + s(1 - 2U_c(z))\hat{\psi}$$

We have the following lemma:

Lemma. If $c > c_{min}$, there is a positive bounded function $\phi(z)$ satisfying $\phi(-\infty) = 0$ and a positive constant μ such that

$$|\hat{\psi}(z, t)| \leq \phi(z)e^{-\mu t} \sup_{\hat{z}}|\hat{\psi}(\hat{z}, 0)/\phi(\hat{z})|$$

for $-\infty < z < \infty$ and $t > 0$.

Given $c > c_{min}$, there are μ and ϕ as described in the lemma. Setting $\hat{\psi} = e^{-\mu t}\phi(z)\Psi$, we have

$$\frac{\partial \Psi}{\partial t} = \frac{\partial^2 \Psi}{\partial z^2} + \left(\frac{2(\partial \phi/\partial z)}{\phi} + c\right)\frac{\partial \Psi}{\partial z}$$

The maximum principle can be applied to this equation to show that

$$|\Psi(z, t)| \leq \sup_{\hat{z}}|\Psi(\hat{z}, 0)|$$

provided that the right-hand side is bounded. Therefore,

$$|\hat{\psi}(x, t)| \leq \phi(x)e^{-\mu t} \sup_{\hat{z}}|\hat{\psi}(\hat{z}, 0)/\phi(\hat{z})|$$

if the initial conditions satisfy

$$\hat{\psi}(z, 0) = O[\phi(z)] \quad \text{as} \quad z \to \pm\infty$$

Let the fundamental solution of this linear problem be denoted by $\Gamma(z, \zeta, t)$. Then

$$\hat{\psi}(z, t) = \int_{-\infty}^{\infty} \Gamma(z, \zeta, t)\hat{\psi}(\zeta, 0) \, d\zeta$$

It follows that

$$\left| \int_{-\infty}^{\infty} \Gamma(z, \zeta, t)\hat{\psi}(\zeta, 0) \, d\zeta \right| \leq \phi(z)e^{-\mu t} \sup_{\hat{z}} |\hat{\psi}(\hat{z}, 0)/\phi(\hat{z})|$$

so we have an estimate of the fundamental solution.

We can rewrite the original nonlinear problem as an integral equation:

$$\psi(z, t) = \int_{-\infty}^{\infty} \Gamma(z, \zeta, t)\hat{\psi}(\zeta, 0) \, d\zeta$$
$$+ \int_0^t \int_{-\infty}^{\infty} \Gamma(z, \zeta, t - t')\mathcal{H}(\psi(\zeta, t')) \, d\zeta \, dt'$$

Taking norms of both sides and using the estimate of Γ, we arrive at the inequality

$$|\psi(z, t)| \leq \phi(z)e^{-\mu t} \sup_{\zeta} |\psi_0(\zeta)/\phi(\zeta)|$$
$$+ \int_0^t \phi(z)e^{-\mu(t-t')} \sup_{\hat{z}} \left| \frac{\mathcal{H}(\psi(\hat{z}, t'))}{\phi(\hat{z})} \right| \, dt'$$

Since $\mathcal{H} = o(|\psi|)$, a standard argument using Gronwall's inequality (see Coddington and Levinson, 1955) shows that there is $\rho > 0$ such that

$$|\psi(z, t)| \leq \phi(z)e^{-\mu t/2} \sup_{\hat{z}} |\psi_0(\hat{z})/\phi(\hat{z})|$$

provided that $\sup |\psi_0(\hat{z})/\phi(\hat{z})| < \rho$. This completes the proof of the following theorem.

Theorem. Let $c > 2\sqrt{s}$. Then there are positive constants ρ and μ and a bounded function $\phi(z)$ satisfying $\phi(-\infty) = 0$ such that the solution of Fisher's equation which satisfies

$$u(x, 0) = \psi_0(x) + U_c(x)$$

can be written as

$$u(x, t) = U_c(x + ct) + \psi(x + ct, t)$$

where U_c is a traveling wave having velocity c and ψ satisfies $\sup_x$ $|\psi(x + ct, t)| \leq Ke^{-\mu t/2}$ for some positive constant K provided $|\psi_0(x)/\phi(x)|$ is bounded.

This theorem shows that any wave that propagates with a speed faster that $c_{\min}$ has a nontrivial domain of attraction. Moreover, the domain of attraction is determined by asymptotic behavior at the leading edge, $z = -\infty$. Such criteria could never be verified for real data. More important, the domain of attraction is determined by behavior near $u \equiv 0$, precisely the region where random sampling effects, which have not been accounted for in the model, are dominant. Therefore, this should be viewed as a negative result.

B. *Wave propagation in oscillatory media*
Oscillatory site models that are coupled by dispersal arise in studies of chemical reactors, spatially distributed ecological systems, and other physical and biological phenomena. The methods presented here are quite similar to averaging methods developed for the study of nonlinear oscillations, and they enjoy the same restrictions encountered there.

First an example is presented that illustrates the method, and then a general formulation of the method is given for a perturbed Hamiltonian system.

Diffusion-coupled van der Pol oscillators. In a hypothetical ecology, a population of herbivores interacts with vegetation in such a way that a stable limit cycle describing the herbivore–vegetation dynamics is established. One of the simple mathematical models that exhibits this phenomenon is the van der Pol oscillator

$$\frac{du}{dt} = \epsilon\left(u - \frac{u^3}{3}\right) + v, \qquad \frac{dv}{dt} = -u$$

where ϵ is a fixed small, positive parameter. If we interpret u and v as (scaled) deviations of herbivore and vegetation biomasses from reference equilibrium values, then we can take this as the site model to describe the dynamics of the population. These site models are distributed on a line, $-\infty < s < \infty$, along which the herbivores migrate. Thus, we consider the system of equations

$$\frac{\partial u}{\partial t} = \frac{\partial^2 u}{\partial s^2} + \epsilon\left(u - \frac{u^3}{3}\right) + v$$

$$\frac{\partial v}{\partial t} = -u$$

for functions $u(s, t)$ and $v(s, t)$, where the diffusion term $\partial^2 u/\partial s^2$ is included to account for herbivore migration.

This system will be analyzed for slowly modulated wave solutions by rescaling the spatial variable with $x = \sqrt{\epsilon}s$. Introducing this as well as a change to polar coordinates given by $u = r\cos\theta$, $v = r\sin\theta$ leads to the equations

$$\frac{\partial r}{\partial t} = \epsilon\cos\theta[\text{diffusion}] - \epsilon r\cos^2\theta\left(\frac{r^2\cos^2\theta}{3} - 1\right)$$

$$\frac{\partial\theta}{\partial t} = -1 - \frac{\epsilon}{r}\sin\theta[\text{diffusion}] + \epsilon\sin\theta\cos\theta\left(\frac{r^2\cos^2\theta}{3} - 1\right)$$

where the diffusion terms are given by

$$[\text{diffusion}] = \frac{\partial^2 r}{\partial x^2}\cos\theta - 2\left(\frac{\partial r}{\partial x}\right)\left(\frac{\partial\theta}{\partial x}\right)\sin\theta$$
$$- r\left(\frac{\partial\theta}{\partial x}\right)^2\cos\theta - r\frac{\partial^2\theta}{\partial x^2}\sin\theta$$

The multitime method derived by Cohen et al. (1977) can be applied to this problem and it gives the following approximations to the solutions:

$$r = r_0(\sigma, x) + O(\epsilon), \qquad \theta = \theta_0(\sigma, x) - t + O(\epsilon)$$

where the error terms are $O(\epsilon)$, at least on the long time interval $0 \leq t \leq T/\epsilon$. The functions r_0 and θ_0 depend on x and the new (slow) time variable $\sigma = \epsilon t$.

Equations for the coefficients r_0 and θ_0 are obtained by averaging the system over the fast time t. This is described in the general formulation given in the next section. The result of averaging is

$$\frac{\partial r_0}{\partial\sigma} = \frac{1}{2}\frac{\partial^2 r_0}{\partial x^2} - \frac{r_0}{2}\left(\frac{\partial\theta_0}{\partial x}\right)^2 - \frac{r_0^3}{8} + \frac{r_0}{2}$$

$$\frac{\partial\theta_0}{\partial\sigma} = \frac{1}{2r_0^2}\frac{\partial}{\partial x}\left(r_0^2\frac{\partial\theta_0}{\partial x}\right)$$

Initial data for these equations would be taken directly from that given for u and v. However, this is a complicated system of equations, and rather than consider the general initial-value problem for it, we will only analyze the system for steady progressing wave solutions.

We introduce the wave variable $\eta = x - \frac{1}{2}c\sigma$ and look for solutions in the form

$$r_0 = 2R(\eta), \qquad \theta_0 = \Theta(\eta)$$

Thus, R and Θ must satisfy ($' \equiv d/d\eta$)

$$-cR' = R'' - R(\Theta')^2 - R^3 + R$$
$$-cR^2\theta' = (R^2\Theta')'$$

The second equation can be integrated, with the result $R^2\Theta' = Ae^{-c\eta}$. Since R and Θ' are to be bounded, we must take $A = 0$, so $\Theta' \equiv 0$. This means that the oscillations are always synchronized in t. (At $R = 0$, Θ' could have discontinuities, but we omit such cases.) With this, the problem reduces to the single problem for the (rescaled) envelope of the population oscillations

$$R'' + cR' - R^3 + R = 0, \qquad R(\pm\infty) \quad \text{bounded}$$

which is easily analyzed by phase-plane methods.

There are three cases of interest.

(a) $c = 0$. This is the case of standing waves, for which there are many solutions to the problem. The phase portrait is given in Figure 5.15, where $Q = R'$ and the arrows indicate the direction of increasing η. The rest points $(\pm 1, 0)$

Figure 5.15. Phase portrait of the envelope of population oscillation $c = 0$.

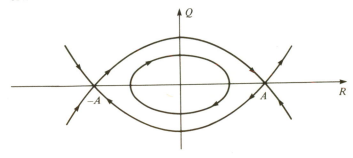

are saddle points, and the separatrices, $Q = \pm(R^2 - 1)/\sqrt{2}$ joining them represent solutions. In addition, (0, 0) is a center and the finite region enclosed by the separatrices is occupied by closed orbits that encircle (0, 0). Each of these orbits is also a solution.

(b) $0 < |c| < 2$. In this case the phase portraits are given in Figure 5.16. Now (0, 0) is a spiral point, stable for $0 < c$ and unstable for $c < 0$, and the separatrices from $(\pm 1, 0)$ spiral into or out of this point. These spirals correspond to oscillations in R and are the only relevant solutions.

Figure 5.16. Phase portrait of the envelope of population oscillation. $0 < |c| < 2$.

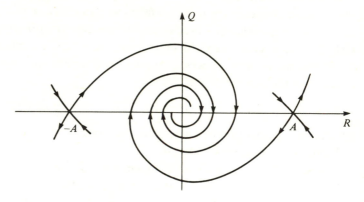

Figure 5.17. Phase portrait of the envelope of population oscillation. $2 \leq |c|$.

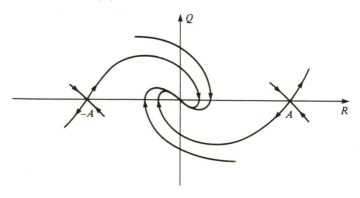

(c) $2 \leq |c|$. In this case, $(0, 0)$ is a node, stable for $2 \leq c$ and unstable for $c \leq -2$, and there are unique trajectories between $(\pm 1, 0)$ and $(0, 0)$. These are the only relevant solutions in this case (See Figure 5.17.)

The analysis just completed shows that the problem has solutions in the form

$$u = 2R(\sqrt{\epsilon}s - \tfrac{1}{2} c\epsilon t) \cos(-t + \delta + O(\epsilon)) + O(\epsilon)$$

$$v = 2R(\sqrt{\epsilon}s - \tfrac{1}{2} c\epsilon t) \sin(-t + \delta + O(\epsilon)) + O(\epsilon)$$

where δ is some fixed constant and R has various forms, depending on the choice of c. The u-component of these solutions is shown for various choices of c in Figure 5.18.

When $c = 0$, the amplitudes of the oscillations are independent of t and only modulated in x. The separatrix solution in Figure 5.18a is significantly modulated only near one point, whereas the closed-orbit solutions in Figure 5.18b lead to a periodically modulated profile. These solutions show that the amplitudes of the solutions are never synchronized in x. For $0 < c < 2$, at fixed x, the solution goes to zero as $t \to -\infty$, but as t increases it undergoes periodic motion with a modulated amplitude corresponding to the passing wave as shown in Figure 5.18c. At some time no further reductions in the amplitude

Figure 5.18. (a) $c = 0$: separatix R.

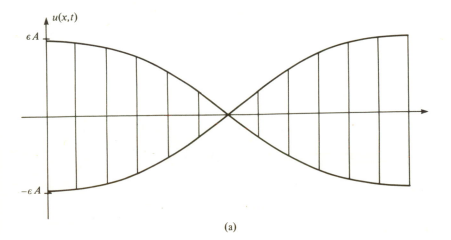

(a)

Figure 5.18. (b) $c = 0$: Periodic R.(c) $0 < |c| < 2$.
(d) $|c| = 2$.

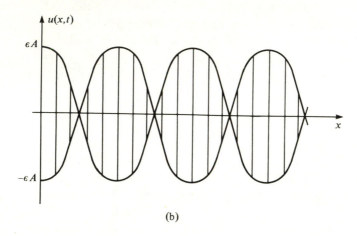

(b)

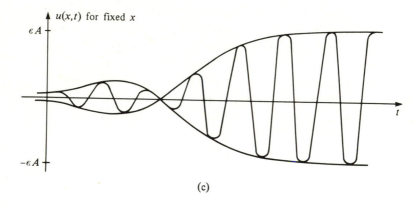

(c)

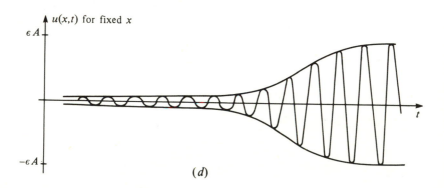

(d)

occur and the steady progressing wave has left behind oscillations that become synchronized asymptotically both in space and time. A similar result is obtained when $2 < c$, as seen in Figure 5.18d, except that the amplitude grows monotonically here. For $c < 0$, the t-axis is reversed in Figure 5.18c and d.

C. *General formulation of the method: Hamiltonian kinetics*

The method used in the preceding subsection can be used for more general diffusion-reaction systems. To illustrate this, we take each site model to be a perturbed Hamiltonian system to which we add diffusion terms

$$\frac{\partial u}{\partial t} = \frac{\partial H}{\partial v} (u, v) + \epsilon f(t, x, u, v; \epsilon) + \epsilon A \frac{\partial^2 u}{\partial x^2}$$

$$\frac{\partial v}{\partial t} = -\frac{\partial H}{\partial u} (u, v) + \epsilon g(t, x, u, v; \epsilon) + \epsilon B \frac{\partial^2 v}{\partial x^2}$$

Here $H(u, v)$, $f(t, x, u, v; \epsilon)$, and $g(t, x, u, v; \epsilon)$ are some smooth functions of their arguments and A and B are some nonnegative constants. The difficulty in using the averaging method for this system is in finding a change of variables that converts the problem into one amenable to the method. This was easily done in the preceding subsection by using polar coordinates because $H = u^2 + v^2$.

There is a general method for finding the correct variables, but it usually cannot be implemented. This proceeds by considering the auxiliary problem

$$\omega \frac{d\tilde{u}}{d\xi} = \frac{\partial H}{\partial v} (\tilde{u}, \tilde{v}), \qquad \omega \frac{d\tilde{v}}{d\xi} = -\frac{\partial H}{\partial u} (\tilde{u}, \tilde{v})$$

Clearly, any solution must satisfy $H(u, v) = \mathcal{H} > 0$, where $\mathcal{H}$ is independent of ξ. Therefore, we can represent a solution of the auxiliary system by

$$\tilde{u} = \tilde{u}(\xi; \mathcal{H}), \qquad \tilde{v} = \tilde{v}(\xi; \mathcal{H})$$

New variables are then introduced into the original problem by

$$u(t, x) = \tilde{u}(\xi, \mathcal{H}), \qquad v(t, x) = \tilde{v}(\xi \mathcal{H}),$$
$$\xi = \xi(t, x,) , \qquad \mathcal{H} = \mathcal{H}(t, x)$$

and the result is the system

$$\frac{\partial \mathcal{H}}{\partial t} = -\epsilon\omega\left(\frac{\partial \tilde{v}}{\partial \xi}\{\cdot\}(f + Au_{xx}) - \frac{\partial \tilde{u}}{\partial \xi}\{\cdot\}(g + Bv_{xx})\right)$$

$$\frac{\partial \xi}{\partial t} = \omega + \epsilon\omega\left(\frac{\partial \tilde{v}}{\partial \mathcal{H}}\{\cdot\}(f + Au_{xx}) - \frac{\partial \tilde{u}}{\partial \mathcal{H}}\{\cdot\}(g + Bv_{xx})\right)$$

where the terms $\{\cdot\}$ must be rewritten in terms of ξ, $\mathcal{H}$ by the change of variables.

The multiscale perturbation method shows that the solution of this problem can be written in the form

$$\mathcal{H} = \mathcal{H}_0(\sigma, x) + O(\epsilon), \qquad \xi = \omega t + \xi_0(\sigma, x) + O(\epsilon)$$

where

$$\frac{\partial \mathcal{H}_0}{\partial t} = -\lim_{T\to\infty}\frac{1}{T}\int_0^T \omega_0\left(\frac{\partial \tilde{v}}{\partial \xi}\cdot(f + Au_{xx}) - \frac{\partial \tilde{u}}{\partial \xi}\cdot(g + Bv_{xx})\right)_0 dt$$

$$\frac{\partial \xi_0}{\partial t} = \lim_{T\to\infty}\frac{1}{T}\int_0^T \omega_0\left(\frac{\partial \tilde{v}}{\partial \mathcal{H}}\cdot(f + Au_{xx}) - \frac{\partial \tilde{u}}{\partial \mathcal{H}}\cdot(g + Bv_{xx})\right)_0 dt$$

where ξ in the integrand is replaced by $\omega t + \xi_0$ at every occurrence. The dependence of ω_0 on t and x is to be chosen so as to ensure that these limits exist. This results in a system of nonlinear partial differential equations for $\mathcal{H}_0$ and ξ_0 that must be analyzed.

A method for analyzing diffusion-reaction systems has been illustrated here through a simple example that exhibits interesting solutions. Although several kinds of steady progressing wave solutions have been described here, their stability has not been investigated. So we have only given a brief description of some possible modes of propagation for the solutions.

The method presented here suffers from two difficulties: determining the appropriate change of variables for the system and analyzing solutions of the averaged system. To date, these have been worked out only for the simplest cases.

Appendix: Derivation of the diffusion approximation
5A.1. *Fokker–Planck equation*

Let α_n be the number of A genes in the nth generation, with transition probabilities given by the Fisher–Wright model with mutation and selection. Recall that the total population size is $2N$. This will be incorporated in the notation by writing $\alpha_n = Y_N(n)$. We define a function of time t by

$$X_N(t) = Y_N([2Nt])$$

where $[x]$ is the greatest integer less than or equal to x. It can be shown that the stochastic process $\{X_N(t)\}$ converges weakly to a diffusion process $X(t)$ as N increases [see Guess (1973)]. The process X has the following properties:

$$E[X(t + dt) - X(t) | X(t) = x] = 2Nm(x)\, dt + o(2N\, dt)$$
$$\mathrm{var}[X(t + dt) - X(t) | X(t) = x] = 2Nv(x)\, dt + o(2N\, dt)$$
$$E[(X(t + dt) - X(t))^j | X(t) = x] = o(2N\, dt) \qquad \text{for } j = 3, 4, \ldots$$

Note that the factors $2N$ arise because of our choice of time scales. Let $\phi(x, t)$ denote the density function of $X(t)$ at time t.

The following is a nonrigorous derivation of an equation for ϕ. Earlier, we had considered the transition probability matrix; now this is replaced by a function q called the *infinitesimal transition probability:*

$$q(dt, x, \xi)\, d\xi = \Pr[\xi \le X(t + dt) - x \le \xi + d\xi | X(t) = x]$$

Therefore,

$$\phi(x, t + dt) = \int \phi(x - \xi, t) q(dt, x - \xi, \xi)\, d\xi + o(dt)$$

Since $\int q(dt, x, \xi)\, d\xi = 1$, we have

$$\phi(x, t) = \int \phi(x, t) q(dt, x, \xi)\, d\xi$$

It follows that

$$\phi(x, t + dt) - \phi(x, t) = \int (\phi(x - \xi, t) q(dt, x - \xi, \xi) \\ - \phi(x, t) q(dt, x, \xi))\, d\xi + o(dt)$$

Next, the integrand is expanded about $\xi = 0$. The result is

$$\phi(x, t + dt) - \phi(x, t) = -\frac{\partial}{\partial x}\left(\phi(x, t) \int \xi q(dt, x, \xi)\, d\xi\right) \\ + \frac{1}{2}\frac{\partial^2}{\partial x^2}\left(\phi(x, t) \int \xi^2 q(dt, x, \xi)\, d\xi + o(dt)\right) + o(dt)$$

Now,

$$\int \xi q(dt, x, \xi)\, d\xi = E[X(t + dt) - X(t) | X(t) = x]$$

and so on, so

$$\phi(x, t + dt) - \phi(x, t)$$
$$= -\frac{\partial}{\partial x}(2Nm(x)\phi(x, t))\ dt + \frac{1}{2}\frac{\partial^2}{\partial x^2}(2Nv(x)\phi(x, t))\ dt + o(2N\ dt)$$

Dividing both sides by dt and passing to the limit $dt = 0$, we have

$$\frac{1}{2N}\frac{\partial\phi}{\partial t} = \frac{1}{2}\frac{\partial^2}{\partial x^2}(v(x)\phi(x, t)) - \frac{\partial}{\partial x}(m(x)\phi(x, t))$$

(Fokker–
Planck
equation)

A rigorous derivation of this is given in Feller (1968). This equation is also referred to as the *forward Kolmogoroff equation*.

$v(x)$ is called the diffusion coefficient and $m(x)$ is the drift of the process. For the Fisher–Wright model with no selection and no mutation, we have for $Z(t) =$ the numer of A genes,

$$m(z) = E[Z(t + dt) - Z(t)\,|\,Z(t) = z] = 2Nz - 2Nz = 0$$
$$v(z) = E[(Z(t + dt) - Z(t))^2\,|\,Z(t) = z]$$
$$= (1/2N)2Nz(1 - z/2N)$$

On the other hand, when we consider the gene frequency at each time step, we set $X(t) = Z(t)/2N$. Then for this process we have

$$m(x) = E[X(t + dt) - X(t)\,|\,X(t) = x] = 0$$
$$v(x) = E[(X(t + dt) - X(t))^2\,|\,X(t) = x] = (1/2N)\,x(1 - x)$$

Therefore, we consider the Fokker–Planck equation for $0 \le x \le 1$. Appropriate boundary conditions for this equation are discussed later.

5A.2. Backward Kolmogoroff equation
The notation will be modified slightly by setting

$$\phi(x, t; x_0) = \Pr[x \le X(t) \le x + dx\,|\,X(0) = x_0]$$

Then using the fact that

$$\phi(x, t + dt; x_0) = \int q(dt, x_0, \xi)\phi(t, x, x_0 + \xi)\ d\xi$$

we repeat the arguments used to derive the Fokker–Planck equation and find that

$$\frac{1}{2N}\frac{\partial \phi}{\partial t} = \frac{1}{2} v(x_0)\frac{\partial^2 \phi}{\partial x_0^2} + m(x_0)\frac{\partial \phi}{\partial x_0}$$

Thus, $\phi(x, t; x_0)$ satisfies both the forward and backward equations, as well as the initial conditions $\phi(x, 0; x_0) = \delta(x - x_0)$.

The approximation $X_N \sim X$ requires that $N \gg 1$. As we originally scaled the time t by a factor $2N$, some care is needed in passing to the limit $dt = 0$ at the same time that N is considered large. The validity of the diffusion approximation is not known in general for arbitrary choices of $m(x)$ and $v(x)$. The case considered next has been established rigorously, however.

5A.3. *Application of the diffusion approximation to the Fisher–Wright model with selection and mutation*

In considering the model with selection and mutation, we found that at each generation the new gene frequencies have a binomial distribution with the expected change being $2N(p|\alpha_n) - \alpha_n$. Therefore,

$$m(x) = v + (1 - \mu - v)\frac{rx^2 + sx(1 - x)}{rx^2 + 2sx(1 - x) + (1 - x)^2} - x$$

$$v(x) = (1/2N)(p|x)(1 - (p|x))$$

If $r = 1 + \rho/2N$, $s = 1 + \sigma/2N$, $\mu = m/2N$, and $v = n/2N$, then

$$m(x) = \frac{n}{N}$$

$$+ \left(1 - \frac{m}{N} - \frac{n}{N}\right)\left[\frac{x + (1/N)(\rho x^2 + \sigma x(1 - x))}{1 + (1/N)(\rho x^2 + 2\sigma x(1 - x))}\right] - x$$

$$= x - x + \frac{1}{2N}[n - x(m + n) + \rho x^2 + \sigma x(1 - x)$$

$$- x(\rho x^2 + 2\sigma x(1 - x))] + O\left(\frac{1}{N^2}\right)$$

$$= \frac{1}{2N}(n - x(m + n)) + \frac{1}{N}(\rho x^2(1 - x)$$

$$+ \sigma x(1 - x)(1 - 2x)) + O\left(\frac{1}{N^2}\right)$$

Similarly,

$$v(x) = (1/2N)x(1 - x) + O(1/N^2)$$

Therefore, the Fokker–Planck equation becomes

$$\frac{\partial \phi}{\partial t} = \frac{1}{2}\frac{\partial^2}{\partial x^2}(x(1-x)\phi) -$$
$$\frac{\partial}{\partial x}\{[x(1-x)(\rho x + \sigma(1-2x)) - mx + n(1-x)\phi]\}$$

for $N \gg 1$. The validity of the diffusion approximation where $r - 1$, $s - 1$, and μ and $\nu = O(1/N)$ has been proved (Guess, 1973). The backward equation is

$$\frac{\partial \phi}{\partial t} = \frac{1}{2}(x_0(1-x_0))\frac{\partial^2 \phi}{\partial x_0^2}$$
$$+ \left\{[x_0(1-x_0)(\rho x_0 + \sigma(1-2x_0)) - mx_0 + n(1-x_0)]\frac{\partial \phi}{\partial x_0}\right\}$$

This can be shown to be the formal adjoint of the forward equation if the boundary values are suitably restricted.

The main case of interest here is the one of absorbing boundaries. In the Fisher–Wright models, the states $\alpha_n = 0$, $\alpha_n = 2N$, once entered, are never left (excluding mutation, of course). They are referred to as absorbing states. Since $\phi(x, t; x_0)$ denotes the probability density function, we see that $\phi(x, t; 0) = 0$ for $x > 0$, $\phi(x, t; 1) = 0$ for $x < 1$. Therefore, we take as boundary conditions

$\phi(x, t; 0) = \phi(x, t; 1) = 0 \quad$ for all $\quad t > 0$
$\phi(x, t; x_0) = $ a smooth function of x at $\quad x = 0$ and $x = 1$

(absorbing boundary)

5A.4. *Stationary distributions: Wright's formula*

The form of the Fokker–Planck equation is reminiscent of a continuity equation where the rate of change of concentration (or the quantity being conserved) equals the negative of the divergence of the flux. In this case,

$$\frac{\partial \phi}{\partial t} = \frac{\partial}{\partial x}\left[\frac{\partial}{\partial x}(v(x)\phi) - m(x)\phi\right]$$

where v and m are defined in preceding sections. The quantity [...] has been interpreted as a probability flux (see Crow and Kimura, 1970; Ludwig, 1974). At equilibrium, the flux should vanish, so

$$(d/dx)(v(x)\phi) = 2m(x)\phi$$

The solution of this equation is

$$\phi(x) = (c/v(x)) \exp\left[2 \int^x (m(x')/v(x')) \, dx' \right]$$
(Wright's formula)

Substituting the value of m and v derived for the Fisher–Wright model, we obtain

$$\phi(x) = cx^{\beta_1-1}(1 - x)^{\beta_2-1} \exp(\alpha_1 x + \alpha_2 x^2)$$

where $\beta_1 = 2n$, $\beta_2 = 2m$, $\alpha_1 = 2\sigma$, $\alpha_2 = \rho - 2\sigma$. Note that β_1 and β_2 vanish when $m = n = 0$. This formula gives the probability density of the equilibrium gene distribution. When $m = n = 0$, there is no mutation to maintain nonfixed states. As we have seen from the martingale calculation (at least for $\rho = \sigma = 0$), fixation is certain. Therefore, there is no regular equilibrium distribution, but everything lies eventually at $x = 0$ and $x = 1$.

Exercises
5.1 (a) Show that the glider (Figure 5.1) configuration is a steady progressing form for the game of life.

(b) Show that the configuration

0	+
0	−

reproduces itself every four steps in the insect model (Section 5.1B).

(c) Determine the distribution of excitation in (b) after 10 generations.

5.2 Let $B(t)$ denote the numbers of bacteria and $N(t)$ the concentration of a limiting nutrient in a growth chamber at time t. The time scale is chosen to correspond to the cell's doubling time when sufficient nutrient is present. The Jacob–Monod model of nutrient uptake is described by $VN/(K + N)$, where V is the maximum nutrient uptake rate per cell and K, called the Michaelis (or saturation) constant, is the value of N at which uptake is at half its maximum rate. The bacteria–nutrient dynamics are described by the equations

$$dB/dt = VY\left(\frac{N}{K + N}\right) B, \qquad \frac{dN}{dt} = -\frac{VN}{K + N} B$$

where Y, called the yield, is the number of cells produced per unit of nutrient taken up.

(a) Show that $VY = \log_e 2$.

(b) Solve the model for N as a function of B.

(c) Substitute this [from part (b)] into the first equation and describe the dynamics of B as a function of t.

5.3 The error function is defined by

$$\mathrm{erf}(x) = \frac{2}{\sqrt{\pi}} \int_0^x e^{-y^2}\, dy \quad \text{and} \quad \mathrm{erfc}(x) = 1 - \mathrm{erf}(x)$$

(a) Show that $\mathrm{erf}(\infty) = 1$ and $\mathrm{erf}(-\infty) = -1$.
(b) Verify that $u(x, t) = \mathrm{erfc}(x/\sqrt{4Dt})/2$ satisfies the diffusion equation $\partial u/\partial t = D(\partial^2 u/\partial x^2)$ and boundary conditions $u(-\infty, t) = 1$, $u(\infty, t) = 0$.
(c) Using the solution in part (b), find the point $x = X(t)$ at which $u(X(t), t) = \frac{1}{2}$. Repeat this calculation for $x = X(t)$ for which $u(X(t), t) = .01$. Sketch the u profile for increasing t.

SOLUTIONS TO
SELECTED EXERCISES

1.1 The Malthus model states that $P_n = r^n P_0$; equivalently,

$$\log P_n = n \log r + \log P_0$$

Set $y_n = \log P_n$, $a = \log r$, and $b = \log P_0$, where $n = 1780$, $1800, \ldots$, and $y_1 = \log(2104)$, $y_2 = \log(2352), \ldots$. The least-squares estimate of a (hence r) is $a = 7.84E - 3$ ($r = 1.008$), respectively. The value predicted by this model is $y_{1920} = 1920a + b = 6023$. The observed population is 5876, and the percent error of this calculation is $(6023 - 5876)/5876 = 2.5\%$.

1.2 The recursion for θ_n can be taken as

$$\theta_{n+1} = 2\theta_n, \qquad 0 < \theta_n < \pi/4$$
$$= \pi - 2\theta_n, \qquad \pi/4 < \theta_n < \pi/2$$

Note that Lebesque measure is invariant for this mapping. In fact, an interval $a < \theta_{n+1} < b$ maps back into two intervals $a/2 < \theta_n < b/2$ and $(\pi - b)/2 < \theta_n < (\pi - a)/2$. The measure (length) of the first is $b - a$ and the total measure of the pre-image intervals is

$$\frac{b - a}{2} + \frac{b - a}{2} = b - a$$

It follows from this fact and the ergodic theorem (see Munroe, 1953) that in the simulation all the cells will have approximately equal number of hits.

1.3 $P_n = (\Pi_{k=0}^{n-1} r_k) P_1$, so for $n = MT$ we have $P_{MT} = (\Pi_{k=0}^{T-1} r_k)^M P_1$. Thus, P_{MT} approaches zero or infinity, respectively as $\Pi_{k=0}^{T-1} r_k$ is less or greater than 1; similarly for P_n. In general, $\log P_n = (\Sigma_{k=0}^{n-1} \log r_n) + \log P_1$. In part (c) we divide by n and pass to the limit $n = \infty$: $\lim (n \to \infty) (1/n) \log P_n = \log \bar{r}$. Thus, for large n, $P_n \sim \bar{r}^n$. If $\bar{r} < 1$, $P_n \to 0$ and if $\bar{r} > 1$, $P_n \to \infty$.

1.4 Let $Q_n = P_n - P^*$. Then

$$Q_{n+1} = f(Q_n + P^*) - f(P^*) = f'(P^*)Q_n + H_n$$

where $H_n = O((Q_n)^2)$. This equation is equivalent to

$$Q_n = (f'(P^*))^n Q_0 + \sum_{k=0}^{n-1} f'(P^*)^{n-1-k} H_k$$

An induction argument shows that if $|Q_0|$ is small, then $Q_n \to 0$.

1.5(a) $P(t) = e^{\rho t} P(0)$.

(b) The change of variables leads to the equations

$$du/dt = \rho u, \qquad dv/dt = \rho u/K$$

Thus, since $(u(0)/v(0)) = P(0)$,

$$P(t) = \frac{KP(0)e^{\rho t}}{(K + P(0))(e^{\rho t} - 1)}$$

(c) $P(t) = \exp(\int_0^t \rho(t')\, dt') P(0)$. Therefore, $\lambda = \lim (t \to \infty)$ $(1/t) \int_0^t \rho(t')\, dt'$ when this limit exists. If ρ is periodic with period T, then $\lambda = (1/T) \int_0^T \rho(t')\, dt'$. If $\rho(t)$ is almost periodic, then λ also exists.

(d) Combine the results of parts (b) and (c).

2.1(a) From Taylor's theorem, we have

$$u(a + h, t + h) = u(a, t) + \left(\frac{\partial u}{\partial a}(a, t) + \frac{\partial u}{\partial t}(a, t) \right)h + O(h^2)$$

Divide this formula by h and pass to the limit $h = 0$.

(b) Introduce a characteristic coordinate s so that $dt/ds = 1$, $da/ds = 1$. Then

$$(\partial u/\partial t) + (\partial u/\partial a) = du/ds$$

First, $t = s$ and $a = s + a_0$, so

$$du/ds = -d(s + a_0, s)u, \quad u|_{s=0} = \mathring{u}(a_0)$$

It follows that $u(s) = \mathring{u}(a_0) \exp(- \int_0^s d(s' + a_0, s')\, ds')$. We eliminate s from this equation ($a_0 = a - t$). Thus,

$$u(a, t) = \mathring{u}(a - t)$$
$$\exp\left(- \int_0^t d(s' + a - t, s')\, ds' \right) \quad \text{for} \quad a > t$$

Next, $t = s + t_0$, $a = s$, so

$$du/ds = -d(s, s + t_0)u, \quad u|_{s=0} = B(t_0)$$

Therefore, $u(s) = B(t_0) \exp(-\int_0^s d(s', s' + t_0)\, ds')$. Eliminating s from this ($s = a$, $t_0 = t - a$), we have

$$u(a, t) = B(t - a)$$
$$\exp\left(- \int_0^a d(s', s' + t - a)\, ds' \right) \quad \text{for} \quad t > a$$

(c) Since $B(t) = \int_0^\infty \beta(a, t)u(a, t)\, da$, we have

$$B(t) = \int_0^t \beta(a, t)B(t - a) \exp\left(- \int_0^a d(s', s' + t - a)\, ds' \right) da$$
$$+ \int_t^\infty \beta(a, t)\mathring{u}(a - t) \exp\left(- \int_0^t d(s' + a - t, s')\, ds' \right) da$$

If $\mathring{u}(a - t)\beta(a, t) = 0$ for $a \geq A$, then

$$f(t) = \int_t^\infty \beta(a, t)\mathring{u}(a - t)$$

$$\exp\left(-\int_0^t d(s' + a - t, s')\, ds'\right) da = 0 \quad \text{for} \quad t > A$$

(d) If d and β depend only on a, then

$$u(a, t) = \mathring{u}(a - t) \exp\left(-\int_{a-t}^a d(s'')\, ds''\right) \quad \text{for} \quad a > t$$

$$= B(t - a) \exp\left(-\int_0^a d(s')\, ds'\right) \quad \text{for} \quad t > a$$

Therefore, $k(a) = \beta(a) \exp(-\int_0^a d(s')\, ds')$. Note that

$$\int_0^\infty \left(\int_0^t k(a)e^{-sa} B(t - a)e^{-s(t-a)}\, da\right) dt$$

$$= \int_0^\infty k(a)e^{-sa}\left(\int_a^\infty B(t - a)e^{-s(t-a)}\, dt\right) da = \tilde{k}(s)\tilde{B}(s)$$

Therefore, $\tilde{B}(s) = \tilde{f}(s)/(1 - \tilde{k}(s))$.

(e) First, we observe that

$$\tilde{k}'(s) = \frac{d}{ds}\tilde{k}(s) = -\int_0^\infty ak(a)e^{-sa}\, da < 0$$

Moreover, $\tilde{k}(\infty) = 0$ and $\tilde{k}(-\infty) = \infty$. Therefore, $\tilde{k}(s)$ crosses the value $\tilde{k} = 1$ exactly once. This root ($s = s_0$) can be found by Newton iteration. Given a value $\tilde{s}_n$, define $\tilde{s}_{n+1}$ by the condition

$$1 - \tilde{k}(\tilde{s}_{n+1}) = 1 - \tilde{k}(\tilde{s}_n) - \tilde{k}'(\tilde{s}_n)(\tilde{s}_{n+1} - \tilde{s}_n) = 0$$

Thus, $\tilde{s}_{n+1} = \tilde{s}_n + (1 - \tilde{k}(\tilde{s}_n))/\tilde{k}'(\tilde{s}_n)$. This sequence $\{\tilde{s}_n\}$ converges to s_0.

(f) Let the roots of $\tilde{k}(s) - 1$ be denoted by $s_0, s_1, s_2, \ldots$, where $s_0 > \operatorname{Re} s_1 \geq \ldots$. If these are simple roots, then the integral breaks up into a sum of integrals,

$$\frac{1}{2\pi i}\int_{s_0+1-i\infty}^{s_0+1+i\infty} e^{st}\frac{\tilde{f}(s)}{1 - \tilde{k}(s)}\, ds = \sum_{m=0}^\infty \frac{1}{2\pi i}\int_{c_m} e^{st}\frac{\tilde{f}(s)}{1 - \tilde{k}(s)}\, ds$$

where c_m is a (small) circle of radius r about the root s_m. Thus, setting $s = s_m + re^{i\theta}$ yields

$$\frac{1}{2\pi i}\int_{\theta=0}^{\theta=2\pi} \frac{\exp(t(s_m + re^{i\theta}))\tilde{f}(s_m + re^{i\theta})ire^{i\theta}\, d\theta}{1 - \tilde{k}(s_m + re^{i\theta})}$$

$$= \frac{1}{2\pi i}\int_0^{2\pi} \left(\frac{\exp(ts_m)\tilde{f}(s_m)}{-\tilde{k}'(s_m)}\right) d\theta + O(r) \to -\frac{\tilde{f}(s_m)}{\tilde{k}'(s_m)}e^{s_m t}$$

$$\text{as} \quad r \to 0$$

Thus, $A_m = -\tilde{f}(s_m)/\tilde{k}'(s_m)$.

(g) Using this formula, we see that

$$\exp(-s_0 t) u(a, t) \rightarrow A_0 \exp\left(-\int_0^a d(s')\, ds'\right) \quad \text{as} \quad t \rightarrow \infty$$

(See Coale (1972) for further analysis of the renewal equation.)

2.2 Verify this directly.

2.3(a) The argument is like that in Exercise 1.4.

(b) Let $G(z) = \sigma z^k + f'(P^*)$ and $F(z) = z^{k+1}$. Use Rouche's theorem from the theory of complex variables to show that all the roots of $F(z) - G(z)$ lie inside the unit circle of the complex plane. It follows that if $Q_{n+1} = \sigma Q_n + f'(P^*) Q_n$, then $Q_n \rightarrow 0$. The argument is completed as in Exercise 1.4.

3.1 Find matrices Z and A, with Z nonsingular and A upper triangular, such that $Z^{-1} P Z = A$. Then the eigenvalues of A are those of P. Let

$$Z = \begin{bmatrix} 1 & 0 & 0 & \cdots & 0 \\ 1 & 1 & 1 & \cdots & 1 \\ 1 & 2 & 2^2 & \cdots & 2^{2N} \\ \cdot & & \cdot & & \\ \cdot & & \cdot & & \\ \cdot & & \cdot & & \\ 1 & 2N & (2N)^2 & \cdots & (2N)^{2N} \end{bmatrix}$$

so $z_{ij} = i^j$. Then the (i, j)th element of (PZ), denoted by $(PZ)_{ij}$, is

$$(PZ)_{ij} = \sum_{l=0}^{2N} p_{il} l^j = E[\alpha_{n+1}^j \mid \alpha_n = i]$$

Also,

$$(ZA)_{ij} = \sum_{l=0}^{j} a_{lj} i^l$$

Therefore,

$$E[\alpha_{n+1}^j \mid \alpha_n = i] = \sum_{l=0}^{j} a_{lj} i^l$$

Using the fact that

$$\sum_{l=0}^{j} a_{lj} i^l = \sum_{l=0}^{2N} p_{il} l^j = E[\alpha_{n+1}^j \mid \alpha_n = i]$$

and properties of the binomial distribution, show that

$$a_{jj} = \frac{2N(2N-1)\ldots(2N-j+1)}{(2N)^j}, \qquad j = 1, \ldots, 2N$$

[see Ewens (1969, pp. 29ff.).]

3.2(a) $(S(n\epsilon + \epsilon) - S(n\epsilon))/\epsilon = -rS(n\epsilon)I(n\epsilon)$. Thus, setting $t = \epsilon n$, we have for $\epsilon \rightarrow 0$,

$$\frac{dS}{dt}(t) = -rS(t)I(t)$$

The other equations follow in the same way.

(b) $dI/dS = -1 + q/rS$. Thus, $I - I_0 = q/r \log_e (S/S_0) - (S - S_0)$.

(c) Since $S \geq 0$ and $dS/dt \leq 0$, S approaches a limit, say $S(\infty)$, as $t \to \infty$. Moreover, since $R \leq S_0 + I_0$ and $dR/dt \geq 0$, R approaches a limit, $R(\infty)$. Thus, $dR/dt \to 0$ as $t \to \infty$, and so $I(t) = (1/q)(dR/dt) \to 0$ as $t \to \infty$. Using this and the formula in (b), we have that

$$S(\infty) = S_0 + I_0 + (q/r) \log_e (S(\infty)/S_0)$$

or equivalently,

$$\frac{S(\infty)}{S_0} = \exp\left\{\frac{r}{qS_0}\left[\left(\frac{S(\infty)}{S_0}\right) - 1 - \frac{I_0}{S_0}\right]\right\}$$

(d) Setting $F = S_\infty/S_0$, we have

$$F - \exp\left(\frac{r}{qS_0}\left(F - 1 - \frac{I_0}{S_0}\right)\right) = 0$$

Solve this using Newton's method [Exercise 2.1(e)]. Other kinds of threshold results are given in Hoppensteadt and Waltman (1970, 1971).

4.1(c) The set of points (x, y, z) in three dimensions for which $x = s^2$ and $z = (1 - s)^2$ defines a parabolic cylinder. The intersection of the plane $x + y + z = 1$ with the first orthant is an equilateral triangle. Its interaction with the cylinder is the set of points $(s^2, 2s(1 - s), (1 - s)^2)$.

(d) These points in triangular coordinates lie on the Hardy–Weinberg equilibrium curve, which is a parabolic curve as in part (c). In the first case $(r > s > t)$ $g_n \to 0$, so the points move to the A_1A_1 vertex. The other cases are similar.

4.2(a) Let $p = (D + H)/P$ and $q = (H + R)/P$. These represent the proportions of the gene pool, which are of type A and a, respectively. An AA fertilized by an A gene produces an AA offspring, and so on. Therefore,

$$dD/dt = bDp + 2Hb(p/2) - d_1D = bPp^2 - d_1D$$
$$d2H/dt = 2bPpq - d_2 2H$$
$$dR/dt = bPq^2 - d_3R$$
$$dP/dt = bP - (d_1D + 2d_2H + d_3R)$$

(b) It follows that

$$\frac{dx}{dt} = \frac{dD/dt}{P} - \frac{D}{P}\frac{dP/dt}{P} = b(p^2 - x) + x(\bar{d} - d_1)$$
$$\frac{dy}{dt} = b(pq - y) + y(\bar{d} - d_2)$$

$$\frac{dz}{dt} = b(q^2 - z) + z(\bar{d} - d_3)$$

$$\frac{dp}{dt} = \frac{d}{dt}(x + y) = p\bar{d} - (d_1 x + d_2 y)$$

where $\bar{d} = d_1 x + 2d_2 y + d_3 z$.

(c) Setting $d_i = d + \epsilon \Delta_i$ for $i = 1, 2, 3$, we have $\bar{d} - d_i = \epsilon(\Delta_1 x + 2\Delta_2 y + \Delta_3 z - \Delta_i) = \epsilon(\bar{\Delta} - \Delta_i)$. Therefore,

$$dx/dt = b(p^2 - x) + \epsilon x(\bar{\Delta} - \Delta_1)$$

$$dy/dt = b(pq - y) + \epsilon y(\bar{\Delta} - \Delta_2)$$

$$dz/dt = b(q^2 - z) + \epsilon z(\bar{\Delta} - \Delta_3)$$

$$dp/dt = \epsilon(p\bar{\Delta} - (x\Delta_1 + y\Delta_2))$$

When $\epsilon = 0$, $dp/dt = 0$ and $x \to p^2$, $y \to pq$, and $z \to q^2$ as t increases. After this initial equilibration to Hardy–Weinberg proportions, p changes slowly according to the equation

$$dp/dt = \epsilon p(1 - p)(\Delta_3 - \Delta_2 - (\Delta_1 + \Delta_3 - 2\Delta_2)p)$$

Depending on the relative sizes of Δ_1, Δ_2, and Δ_3, p approaches 0, 1, or $(\Delta_3 - \Delta_2)/(\Delta_1 + \Delta_3 - 2\Delta_2)$ as $\epsilon t \to \infty$.

4.3(a) Since $r^2 = x^2 + u^2$,

$$r\frac{dr}{dt} = x\frac{dx}{dt} + u\frac{du}{dt} = \mu r \cos\theta\left(r\cos\theta - r^3\frac{\cos^3\theta}{3}\right)$$

Therefore,

$$\frac{dr}{dt} = \mu r\left(\cos^2\theta - r^2\frac{\cos^4\theta}{3}\right)$$

Since $\theta = \tan^{-1}(u/x)$,

$$r^2\frac{d\theta}{dt} = x\frac{du}{dt} - u\frac{dx}{dt} = r^2 - r\sin\theta\left(r\cos\theta - r^3\frac{\cos^3\theta}{3}\right)$$

Therefore,

$$\frac{d\theta}{dt} = 1 - \mu\sin\theta\cos\theta\left(1 - \frac{r^2\cos^2\theta}{3}\right)$$

(b) $d\bar{r}/dt = (\mu\bar{r}/2)(1 - \bar{r}^2/4)$, $d\bar{\theta}/dt = 1$. Thus, if $\bar{r}(0) > 0$, $\bar{r} \to 2$ as $t \to \infty$. Note that this occurs on the slow time scale μt. Thus, the point $(x(t), u(t))$ approaches a periodic orbit that has amplitude 2 as t increases.

5.1(a) When sufficient nutrient is present ($N \gg 1$), then $dB/dt = VYB$ or $B(t) = \exp(YVt)B(0)$. Since the time scale corresponds to cell doubling, $B(1) = 2B(0)$ or $\exp(VY) = 2$.

(b) Since $dN/dB = -1/Y$, $N = N(0) - (B - B(0))/Y$.

(c) $\dfrac{dB}{dt} = VY\left(\dfrac{YN(0) - B + B(0)}{YK + YN(0) + B(0) - B}\right) B$. Therefore, $B(t) \rightarrow$ $N(0) + B(0)/Y$.

5.3(c) If $\text{erfc}(X(t)/\sqrt{4\,Dt}) = \text{constant}$, then $X(t) = c_0 (\sqrt{4\,Dt})^{1/2}$. If the constant is $1/2$, then $c_0 = 0$. If the constant is small, then c_0 is very large.

REFERENCES

Aronson, D., 1976. Topics in Nonlinear Diffusion, *CBMS*, SIAM, Philadelphia.

Bailey, N. T. J., 1957. *The Mathematical Theory of Epidemics*. Charles Griffin, London.

Berkeley *Symposium on Probability and Mathematical Statistics* (6th)(1972). Vol. 5. University of California Press, Berkeley, Calif.

Bernardelli, H., 1942. Population waves. *J. Burma Res. Soc. 31:* 1–18.

Beverton, R. J. H., and S. J. Holt, 1957. On the dynamics of exploited fish populations. Ministry of Agriculture, Fisheries and Food (London), *Fish Investment Ser.* 2(19).

Cannings, C., 1974. The latent roots of certain Markov chains arising in genetics. *Adv. Appl. Prob. 6:* 260–90.

Canosa, J., 1973. On a nonlinear equation of evolution. *IBM J. Res. Dev. 17:* 307–13.

Cavalli-Sforza, L. L., and W. A. Bodmer, 1971. *The Genetics of Human Populations*. W. H. Freeman, San Francisco.

Charlesworth, B., 1980. *Evolution in Age Structured Populations*. Cambridge University Press.

Clark, C. W., 1976. *Mathematical Bioeconomics: The Optimal Control of Renewable Resources*. Wiley-Interscience, New York.

Coale, A. J., 1972. *The Growth and Structure of Human Populations: A Mathematical Investigation*. Princeton University Press, Princeton, N.J.

Coddington, E., and N. Levinson, 1955. *Theory of Ordinary Differential Equations*. McGraw-Hill, New York.

Cohen, D. S., F. C. Hoppensteadt, and R. M. Miura, 1977. Slowly modulated diffusion processes. *SIAM J. Appl. Math. 33*(2): 217–29.

Conway, J. H., 1970. Mathematical games. *Sci. Am.,* October, pp. 120–3.

Crow, J. F., and M. Kimura 1970. *An Introduction to Population Genetics Theory*. Harper & Row, New York.

Euler, L., 1760. Recherches générales sur la mortalité et la multiplication du genre humain. *Mem. Acad. R. Sci. Belles Lett. (Belg.) 16:* 144–6.

Ewens, W. J., 1969. *Population Genetics*. Methuen, London.

Fibonacci, L., 1202. Tipographia. delle Scienze Mathematichee Fisiche, Roma.

Feller, W., 1966, 1968. *An Introduction to Probability Theory and Its Applications,* vols. 1, 2. Wiley, New York.

Fisher, R. A., 1930. *The Genetical Theory of Natural Selection*. Dover, New York.

—1936. The wave of advance of an advantageous gene. *Ann. Eugen. 7:* 335–69.

Fleming, W., 1975. A selection–migration model in population genetics. *J. Math. Biol. 2:* 219–33.

Fourier, J., 1822. *Analytical Theory of Heat*. Dover, New York.

Greenberg, J. M., B. D. Hassard, and S. Hastings, 1978. Pattern formation and periodic structures in systems modelled by reaction–diffusion equations. *Bull. AMS 84:* 1296–327.

Guess, H. A., 1973. On the weak convergence of Wright–Fisher models. *Stoch. Proc. Appl. 1:* 282–306.

Gulland, G, 1974. *The Management of Marine Fisheries.* University of Washington Press, Seattle.

Hammersley, J. M., and D. C. Handscomb, 1964. *Monte Carlo Methods.* Methuen, London.

Harris, T. E., 1963. *The Theory of Branching Processes.* Springer-Verlag, New York.

Holling, C. S., 1965. The functional response of predators to prey density and its role in mimicry and population regulation. Mem. Entomol. Soc.

Hoppensteadt, F. C., 1975a. Analysis of a stable polymorphism. *J. Math. Biol. 2:* 235–40.

— 1975b. Mathematical Theories of Populations: Demographics, Genetics and Epidemics. *CBMS,* vol. 20. SIAM, Philadelphia.

— and N. Gordon, 1975. Nonlinear stability analysis of static states which arise through bifurcation. *Commun. Pure Appl. Math. 28:* 355–73.

— and J. M. Hyman, 1977. Periodic solutions to a discrete logistic equation. *SIAM J. Appl. Math. 32:* 73–81.

— and W. Jager, 1980. Pattern formation by bacteria. *Lect. Notes Biomath. 38:* 68–81, Springer Verlag.

— and J. B. Keller, 1976. Synchronization of periodical cicada emergences. *Science 194:* 335–7.

— and W. Miranker, 1977. Multi-time methods for difference equations. *Stud. Appl. Math. 56:* 273–89.

— 1979. Slow selection analysis of genetic traits in synchronized populations. *Rocky Mtn. J. Math. 9:* 93–7.

— and I. Sohn, 1980. A multiple species fishery model. NATO Symposium. *Appl. Oper. Res. 1.*

— and P. Waltman, 1970, 1971. A problem in the theory of epidemics. *Math Biosci. 9:* 71–91; *10:* 133–45.

Isaacson, D. L., and R. W. Madsen 1976. *Markov Chains: Theory and Applications.* Wiley, New York.

Jolly, C., and F. L. Brett 1973. *J. Med. Primatol.*

Karlin, S., 1966. *A First Course in Stochastic Processes.* Academic Press, New York.

— 1972. Some mathematical models in population genetics. *Am. Math. Monthly,* 699–739.

Kermack, W. O., and A. G. McKendrick, 1927, 1932. . . . A contribution to the theory of epidemics, I, II. . . . *Proc. R. Soc. Lond.* Sec. A *115:* 700–21; *139:* 55–83. . . .

Keyfitz, N. and B., 1970. Translation of Euler's population paper. *Theo. Pop. Biol.* 1: 307–14.

Keyfitz, N., and W. Flieger, 1971. *Population: Fact and Methods of Demography.* W. H. Freeman, San Francisco.

Kolmogoroff, A., I. Petrovsky, and N. Piscounov, 1936. *Mosc. Univ. Bull. Math. Ser. Int. A1:* 1–25.

Leslie, P. H., 1945, 1948. *Biometrika 33:* 183–212; *35:* 213–43.

Lotka, A..J., 1922. The stability of the normal age distribution, *Proc. Natl. Acad. Sci. 8:* 339–45.

Ludwig, D. A., 1974. Stochastic population theories. *Lect. Notes Biomath.* vol. 3, Springer-Verlag.

— and B. Haycock, 1976. MacDonald's work on helminth infections. *Case stud. Applied Math., CUPM,* pp. 313–43.

MacDonald, G., 1965. The dynamics of Helminth infections with special reference to schistosomiasis. *Trans. R. Soc. Trop. Med. Hyg. 59:* 489–506.

McKendrick, A. G., 1926. Applications of mathematics to medical problems. *Proc. Edinb. Math. Soc. 44:* 98–130.

McKusick, V. A., 1969. *Human Genetics.* 2nd ed. Prentice-Hall, Englewood Cliffs, N.J.

Malthus, T. R., 1798. *An essay on the principles of population.* London.

May, R. M., 1973. *Stability and Complexity of Model Ecosystems.* Princeton University Press, Princeton, N.J.

Mendel, G. J., 1865. Versuche über Pflanzen-Hybriden. *Verh. Naturforsch. Ver. Brunn. 10.*

Moe, G., W. Rheinboldt, and J. Abildskov 1964. A computer model of atrial fibrillation. *Am. Heart J. 67:* 200–20.

Moran, P. A. P., 1962. *The Statistical Processes of Evolutionary Theory.* Clarendon Press, Oxford.

Munroe, M. E., 1953. *Introduction to Measure and Integration.* Addison-Wesley, Reading, Mass.

Nasell, I., and W. M. Hirsch 1973. The transmission dynamics of schistosomiasis. *Commun. Pure Appl. Math. 26:* 395–453.

Novick, R., and F. C. Hoppensteadt 1978. On plasmid incompatibility. *Plasmid 1:* 421–34.

Ricker, W. E., 1954. Stock and recruitment. *J. Fish. Res. Board Can. 14:* 669–81.

Sarkovski, A. N., 1964. *Ukr. Mat. Zh. 16:* 61–71.

Simons, C., 1979. Debut of the seventeen-year-old cicada. *Nat. Hist. Mag.,* May 1979.

Thom, R., 1975. *Structural Stability and Morphogenesis.* W. A. Benjamin, Reading, Mass.

Vainberg, M. M., and V. A. Trenogin 1974. *Theory of Branching of Solutions of Nonlinear Equations.* Noordhoff, Leyden.

Verhulst, P. F., 1845. Recherches matématiques sur la loi d'accroissement de la population. *Mem. Acad. Roy.,* Belgium, 18: 1–38.

Wiener, N., and A. Rosenbleuth 1946. The mathematical formulation of the problem of conduction of impulses in a network of connected excitable elements. *Arch. Inst. Cardiol. Mex. 16:* 205–65.

Wright, S., 1931. Evolution of Mendelian genetics. *Genetics 16:* 97–159.

AUTHOR INDEX

145

SUBJECT INDEX

ABO blood system, 81
absorbing boundary, 132
absorbing states, 59, 64, 132
age distribution vector, 35
algebraic iteration, 2, 4, 32, 36, 41, 50,
    63, 69, 79, 82, 83, 86, 96
allele, 47
amplification rate, 97
auxotrophs, 102
averaging, the method of, 78, 80, 91,
    128

bacterial genetics, 47
balance equation (input–output), 22,
    24
balanced state, 11
bamboo, 16
below-the-line items, 22
Bernardelli population waves, 44
Bessel function, 87, 106
Beverton–Holt fishery, 18
bifurcation, 68, 85, 89
bifurcation diagram, 8, 88
binomial coefficient, 51
binomial distribution, 62, 70, 131
bionomic equilibrium, 19
birth–death process, 73
branching process, 65
buffer, 101
by-catch, 27

carrying capacity, 30
census, 29, 37
central limit theorem, 97
chaos, 7
characteristic equation, 40, 46
characteristics, method of, 74, 136
chromosome, 47
cicadas, 12
clamdigger's itch, 85
cobwebbing, 3
cod–haddock fishery, 23
coefficient of catchability, 17

copy number, 51
crystal test, 101

DeFinetti diagram, 89
demand vector, 22
depensatory reproduction curve, 7
diffusion equation, 96
    coefficient, 130
    as markov chain approximation, 96,
        128
    in radial coordinates, 103, 122
diffusive instability, 99
diploid, 60
Dirac delta function, 115, 131
discounted future revenues, 20
dishonest matrices, 42
dispersal processes, 92
dispersion relation, 97
disruptive selection, 76
DNA, 47
domain of attraction, 11, 121
doubly stochastic matrix, 58
drug resistance, 52
dynamical systems, 127

ecology matrix, 24
effort–harvest relation, 17
effort–yield diagram, 18
eigenvalue, 34, 41, 64, 100, 138
epidemic models, 69
ergodic theorem, 135
error function, 134
Euler approximation, 82
exchange of stabilities, 9
excitable media, 95
exhaustible resource, 16
extinction probability, 67, 74

Fibonacci renewal equation, 31, 37
Fibonacci reproduction matrix, 33, 41
Fibonacci sequence, 31, 36, 75
final size of epidemic, 72, 73, 84
fisheries, 16

147